LAUGHTER
on the
WING

Volume 1

Hello Collin,

I look forward to meeting you on your next visit down under. Untill then a little light reading for you.

Best Wishes,
Kurt Kranen
BRIS AUSTRALIA.

INTRODUCTION

Welcome to *Laughter on the Wing* Volume 1, a selection of the funniest anecdotes and stories from *Australian Aviation* magazine's *Tales* and *On the Airbands* columns.

For more than a decade the most popular column in *Australian Aviation* has been Bob Bell's *On the Airbands*. Over the years spies from around Australia and the world have sent in humorous transmissions they have heard between aircraft and air traffic controllers. We went digging through dozens of our old magazines searching for the best tales featured in *On the Airbands* to print the most humorous gems here. Some were long forgotten, but others are timeless classics.

The *Tales* column in AA was an instant hit when introduced a few years ago. Almost everyone who has been involved with aviation in some form or other for any length of time can recall mirthful incidents, faux pas, near misses and planned practical jokes. These Tales are often recounted over the bar late at night and some are part of aviation folklore. Hopefully we've captured some of the better stories here for all to read for posterity. As far as we know, most of the Tales are at least based on the truth, but then again why let truth get in the way of a good story!

We have also selected a number of humorous items from the internet that readers may not have already seen, and while they are not specific aviation related, we certainly got a kick out of them and we hope you do too.

This book was the result of the labours of a number of people. Without Bob Bell there would be no *On the Airbands* column, while credit for the stories plucked from *Tales* goes to those *Australian Aviation* readers who took the time to write in and share their humorous and often embarrassing aeronautical experiences. Ian Hewitt was responsible for editing much of the magazine copy into book form, while Sharon O'Neill produced the cover artwork.

We hope you enjoy their efforts, and look out for *Laughter on the Wing* Volume 2.

Jim Thorn
Managing Editor and Publisher
Australian Aviation
Canberra, July 1997

Published by Aerospace Publications Pty Ltd (ACN: 001 570 458), PO Box 1777, Fyshwick ACT 2609, publishers of monthly *Australian Aviation* magazine. http:www.ausaviation.com.au
Production Manager: Maria Davey

ISBN 1 875671 29 3

Proudly printed in Australia by Pirie Printers Pty Ltd, 140 Gladstone Street, Fyshwick 2609
Distribution in North America by Motorbooks International, 729 Prospect Ave, Osceola, Wisconsin 54020, USA, fax: (715) 294 4448. Distribution throughout Europe and the UK by Airlife Publishing Ltd, 101 Longden Rd, Shrewsbury SY3 9EB, Shropshire, UK, fax (743) 23 2944.

I heard a good story the other night from an old friend who during the early eighties was seconded to the USAF flying F-15 Eagles out of Luke AFB, Arizona.

It seems my friend and the squadron commander were ferrying their F-15s from Luke across the continental USA to Langley AFB, Virginia. The weather that day was fairly unseasonal with a fair degree of turbulence around the 20,000ft to 40,000ft levels. Over the radios came a never ending succession of airline pilots asking one another whether it was any less turbulent at higher altitudes etc. This went on for more than an hour while the two Eagle pilots smoothly cruised at FL50 well above the aluminium traffic jam.

With a slight touch of smugness, the F-15 commander came on the airways with the following gem: “To all stations, we’re a couple of Eagles up here at fifty thousand feet, would ya like to join us?” The Eagle pilots soon had the smiles wiped off their faces though as, after about a 10 second silence, an unidentified airline captain smoothly replied – “Hi ya Eagles, up there at fifty thousand feet, we appreciate the air is real smooth up there but how is ya paycheck!”

The final scene has the two Eagles trucking on silently towards Langley without a further word being said.

F-15 Eagles

MDC

~ ~ ~ ~

The aircraft carrier HMAS *Melbourne* was returning home after a long cruise through South East Asia where the crew had wisely taken the opportunity to refresh their liquor stocks.

Now, we won’t suggest for a moment that anybody was really smuggling more than their legal allocation of alcoholic beverage into Australia, but it did come to pass that occasionally this might just occur. Legend has it that at this point in time one of the carrier’s Gannets would have its bomb bay stockpiled with something more akin to liquid explosive than mere torpedoes, bombs or mines.

With the carrier heading towards Garden Island, the air group launches for its short flight to Naval Air Station, Nowra. As befits such an occasion each squadron overflies the base in tight formation much to the approval of the ground crews, family and assorted top brass.

Now, if you go too fast in a Gannet, one way of slowing the beast down was to momentarily open the bomb bay doors to increase drag. Thus our Gannets are approaching Nowra a little too fast and are rapidly closing on the formation ahead of them and you guessed it, the CO quickly opens his bomb bay doors to wash off speed.

Meanwhile, our hapless friend (who is nowadays somebody of note in the aviation industry and will remain anonymous) hits the door operating lever to avoid hurtling ahead of the formation. A split second after opening the doors he realises what he has done as the entire squadron silently watches thousands of dollars of bottled booty hurtle towards Jervis Bay, where it remains to this day!

We certainly know who the drinks were on in the wardroom that night!

Fairey Gannet

Defence PR

~ ~ ~ ~

Sometimes it just never fails to amaze what comes across our desk via press releases. Take this example describing a Jet Provost rebuild.

"The aircraft was constructed in 1963 by BAC, then purchased by South Australian Federation (Yemen) as a ground attack aircraft to support the spread of communications into the United Arab Emirates. The aircraft was then returned to BAC after three years and 400 flying hours due to insufficient technical support and skilled pilots. It was then completely updated and refurbished by BAC and purchased by the Republic of Singapore Air Force as a ground attack/trainer aircraft. After four years the RSAF retired the Provost from service and replaced with the Stickmaster aircraft."

Okay, why are South Australians so interested in supporting communication in Yemen? If the Provost was so unreliable how did it accumulate a pretty healthy 400hrs combat time in such a rugged environment? And just what is a Stickmaster?

~ ~ ~ ~

I well remember my first solo cross country in a Tiger Moth.

It took an hour and a quarter and the routing was Point Cook/Werribee/Meredith/Lethbridge and home.

Sensing that I was a little nervous at setting off solo on such a great journey, my instructor tried to relax me with a little humour. His advice was that if I got lost I should take my watch off, whirl it around my head, and let it go. It would "go west" – I was then to fly in the opposite direction which would take me back to Point Cook!

~ ~ ~ ~

This story was overheard while visiting the US Navy test station at Patuxent River during the early days of the F/A-18 program.

One of the Hornet test pilots (who was on exchange to Navy) had cut his teeth on the much revered Century Series of fighters in his younger years and had in turn always been a member of what USAF respectfully calls the 'fighter community'.

While visiting a Strategic Air Command base in the fifties in his F-104 Starfighter he ran into one of his training course buddies who had gone on to great things with SAC. After looking over the rocket-like Starfighter, his bomber buddy suggested that he experience the thrill of flying in a B-47 Stratojet, the pride of SAC.

Understandably our Starfighter driver was never a big fan of such underpowered giants, but not wanting to disappoint his friend he decided that he could only learn from the experience so he gladly accepted. A B-47 was duly rolled out to the flightline with our jet jockey firmly ensconced in the rear seat. Being a desert base the day was already hot as the Stratojet taxied out to the 4000m long strip of asphalt. At the threshold the six engined beast was run up to full power amid a cacophony of noise and smoke that truly left the Starfighter driver in awe. This could be exciting!

The experience of being pinned in the seat by brute force on brake release however was not to eventuate as the silver giant sedately trundled down the long runway picking up speed at the apparent rate of a steam train. Understandably, our fighter jock began to get agitated as the 5000ft marker drifted by with less than 100kt (about 185km/h) on the dial and little apparent acceleration left in the beast. The story goes that by this stage our fighter jock felt so insecure that he had mentally prepared to eject fearing a complete overrun, when finally the Stratojet stumbled into the sky.

Apparently the big jet handled appropriately when in its natural element at 30,000ft though the way it got there taught our fighter friend to think twice before accepting another ride in one of SAC's multi engined giants.

USAF

B-47 Stratojet

~ ~ ~ ~

We don't really know if this tale is true though its source assures us it is. You can be the judge.

It is late in the afternoon and HMAS *Melbourne* is departing on a routine training exercise with a number of new crew members. One junior officer has apparently imbued too many drinks in his brief but fun filled stay in Sydney and decides to sleep if off before the coming of the dawn and duty time.

Finding his cabin empty, he decides on the top bunk and begins to clamber up the small ladder when the ship begins to roll in the increasing swell. Well, this lad is not on the tall side, damn small in fact, and he slips off the ladder into one of the very large drawers that were a hallmark of *Melbourne.*

Another interesting design aspect of these drawers is that they had large flip down clips which were used to secure them in anything other than a calm sea state. Get the picture?

Finding that the drawer is filled with soft clothing our hapless friend settles into a deep sleep, not really caring that he didn't make his bunk. Anyway, *Melbourne* turns suddenly with a roll that propels the drawer shut with a force that throws the clip down to lock the drawer securely in place.

Aircraft carriers of course are noisy places at the best of times and despite the thumping and screaming of our young officer, nobody heeds his dilemma until the next morning when the steward makes a routine visit to the cabin. Not believing what he is seeing, the steward liberates our friend, who immediately takes on a reputation throughout the senior service that few would want to lay claim to!

~ ~ ~ ~

Back in the mid eighties the then Secretary of the (US) Navy, John Lehman, was known as an avid warbird enthusiast. So much so that he had his own immaculately restored Vought F4U Corsair which he flew almost every Sunday out of Andrews AFB, near Washington DC.

Our story has the Number One man in the USN approaching Andrews after his morning sortie and expecting his usual VIP expedited entry instantly approved from an always obliging controller. Upon calling the tower however he is told that his is second in line behind a USAF C-141 Starlifter which is already established on finals.

"Tell those guys to orbit and clear me immediately for a straight in approach" came the quick and *very* authoritative response from the Navy supremo. Before the tower could reply though this more than amusing response filled the capital's airwaves. "This (in a feminine southern drawl) is USAF Military Airlift Command Starlifter to Corsair, this aircraft is on finals and cleared to land and please note that there are *no* guys on this flightdeck, this *sir* is an *all* female crew"!

Legend has it that the jet transport landed as planned with a very subdued Secretary of the Navy taking his second place in the queue without further comment.

F4U Corsair

Paul Merritt

~ ~ ~ ~

A group of young aviators had been detailed to exchange pleasantries with the Chief of Naval Staff, who had just completed an inspection of the carrier. In the manner of these occasions they were grouped around the great man on the flightdeck, trying to think of what to say if they should be asked a question.

Now it so happened that CNS had been in the Navy as a midshipman during the First World War and somehow managed to scrounge a ride in an observation balloon during a visit to the front.

To demonstrate his prowess in matters aeronautical, he mentioned his ride and enquired to the audience at large as to whether anyone had ever flown a balloon?

A course mate of mine who had decided that his future lay in civil aviation rather than the military volunteered that he had indeed flown a balloon.

When a somewhat surprised CNS enquired as to the occasion and the details, my friend replied "Oh, only on a piece of string sir".

~ ~ ~ ~

Following a bombing mission over South Vietnam, a B-52 is homeward bound for Guam, several thousand klics to the east.

As it trundles on, its eight engines humming reassuringly in the cold air, two loitering F-4 Phantoms formate on the giant bomber to 'escort' it from Vietnamese airspace.

After a little friendly banter between the crews of the fighters and the giant bomber the Phantom lead serenely performs a roll over, around and under the lumbering giant to settle nicely off its starboard wingtip.

"Very impressive, very smooth" comes the response from the Stratofortress skipper.

"Thank you", exclaims an obviously superior feeling Phantom pilot. "Now, why don't you show us what you can do?" suggests the Phantom pilot in a semi mocking tone amidst subdued laughter from his backseater.

Well, nothing much happens for about three minutes and the fighter pilots are wondering what's going on.

"Well what did you do?" asks a curious Phantom lead.

"Well son, we did something that you boys could never do, we just shut down two engines!"

Boeing B-52 Stratofortress

~ ~ ~ ~

Fokker F27 Friendship

Many years ago with an Australian regional airline, it was the practice to play tricks on newly graduated and totally unsuspecting cabin crew.

One such event occurred on a Melbourne/Canberra/Melbourne service involving an overnight stay in Canberra. The aircraft used at the time was the Fokker F27 Friendship. The day began with crew arriving at the airport at 5.30am with the temperature at minus 4 degrees. A very heavy frost had caused the top surface of the wings to ice up. The new flight attendant was then told that her job would be to scrape the ice off the wings as we all know that aircraft can't fly iced up.

She was given the appropriate instrument with which to do this and the ground staff, being duly informed of the joke, obliged by placing a set of steps for the FA to reach the wing. She genuinely believed this had to be done and it wasn't until she was about to make the first scrape that the first officer told her that her services were not required!

Fortunately, she accepted the joke and had a good laugh at her innocence, and continued with her real duties.

~ ~ ~ ~

Back in the early eighties a businessman travelled frequently from Adelaide to Melbourne on an Australian airline which shall remain nameless.

His usual procedure involved departing Adelaide around 6.45am, visiting associates in and around Melbourne during the day, then returning to Adelaide in the evening.

On this particular day, our friend had achieved some considerable business success and celebrated over a lengthy business lunch. Needless to say, he was in a rather mellow mood when departing Melbourne that evening on a Melbourne/Adelaide/Perth flight, and following a few more noggins in flight promptly fell asleep.

When the aircraft commenced its descent into Adelaide, our friend was awakened by a Flight Attendant with a request to fasten his seat belt. This he did before promptly falling asleep again.

When he awoke, his mind still befuddled, he was puzzled to see the cabin crew serving refreshments and immediately queried the aforementioned FA regarding their intended arrival time in Adelaide.

The FA nearly dropped her coffee pot as the true situation dawned on her because at that time the aircraft was an hour out of Adelaide enroute to Perth!

It would appear that the cabin crew failed to properly crosscheck the passenger list prior to departing Adelaide, opting for a head count only, and because one passenger had missed the flight for the Adelaide/Perth leg, the numbers checked out OK.

What followed was a series of frantic radio messages from the aircraft to Perth, and on arrival there, our friend was entertained by a very apologetic airport manager of the nameless airline. The wayward traveller even managed to maintain a sense of humour with a phone call to his boss in Adelaide (around 2.00am Adelaide time) asking if there was any company business he could attend to while he was in Perth!

Arrangements were quickly completed to book him on the first available flight out of Perth and this happened to be the 'Red Eye Special' departing around 1.00am. Unfortunately, this flight proceeded directly from Perth to Melbourne, and as the summer sun rose in the morning our friend was overflying his Adelaide home at 30,000 plus feet!

Back in Melbourne again and somewhat worse for wear, he was directed onto the next flight to Adelaide finally arriving home around 10.30am.

In all, his journey took some 15 hours to complete compared to the normal Melbourne/Adelaide flight time of around one hour. Total distance travelled – approximately 6300km!

Needless to say, embarkation procedures of the particular airline were smartened up after that little episode!

~ ~ ~ ~

This story was told to us by an old school teacher who is also a private pilot.

During an exchange visit to the PRC he had decided to utilise a Chinese airline as transport to his rural destination. A few hours into the flight, the copilot felt the call of nature and excused himself from the flightdeck to visit the toilet. He got up, closed the flightdeck door behind him and proceeded to the rear of the aircraft.

Sometime after, he finished his duties and made his way back to the cockpit. The copilot turned the door knob but it would not budge! He tried rattling it a few times but the door still would not open. The normal thing to do would be to knock on the door and the copilot did this and luckily enough, the pilot opened the door for him from the inside of the cockpit.

Having plenty of time, the two flightdeck crew decided to have a look at what was wrong with the door. They engaged the autopilot and went outside to the passenger cabin, closing the door behind them.

The passengers soon figured out the problem and got scared when they heard the two crew banging the door again and trying to pick the lock to no avail. After some more attempts they got the fire axe and had to bang the door down and deform it so that it looked like how the aircraft might look if they didn't open it in time!

The aircraft eventually landed safely at its destination, albeit with two very embarrassed crewmembers at the controls.

~ ~ ~ ~

Our scene is a very dark though calm winter night over the North Atlantic several hundred klics to the west of Ireland.

A then Strategic Air Command B-52 is looking for its KC-135 tanker link up in a no comms, no lights maximum security rendezvous. Obviously something has gong wrong as the 52 cannot locate the tanker anywhere. Finally out of the gloom the bomber crew notice a faint silhouette of a large jet shape ahead and above them about where the tanker would be expected to be.

Success at last! The eight engined B-52 slowly formates immediately behind and below the Stratotanker still in its lights-out no-comms configuration and awaits the flashing of coloured signal lamps from the stern refuelling window of the KC-135 and the lowering of the refuelling boom so as to be able to take on fuel.

Minutes pass as the two jets fly in tight company through the black night and nothing happens. No torchlight code from the back of the Stratotanker and definitely no lowering of the refuelling boom. Finally in frustration, with the B-52 crew signalling frantically all the while, the B-52 skipper decides to break formation and fly alongside the tanker to see if they can get a response from the flightdeck.

Silence ensues however as the giant bomber comes up alongside a commercial 707 passenger jet with probably 160 dozing passengers oblivious to the close company they have just been keeping!

Bill Lines

A B-52 prepares to take on fuel

~ ~ ~ ~

Who said the RAAF takes life altogether too seriously?

Not according to a story concerning a well known aerobatic instructor based at the Sydney general aviation airport of Bankstown.

Prior to retiring from the RAAF a few years back, this flier held the job of transport operations staff officer at the then Operational Command, responsible for the tasking of all RAAF transport aircraft. One day its seems, RAAF Richmond had a requirement to carry a toy poodle as part of a personnel shift.

"Because dogs are full of piddle and poo, which don't go very well with aircraft structures, getting approval to carry dogs on RAAF aircraft, even toy poodles, was very difficult," recounted a former 37 Sqn (C-130E) executive. "But we thought we'd give it a try and so a signal went off to our friend, who was the authority on these things."

A few days later back came the reply from the august and sober corridors of Operational Command: "Reference signal requesting approval for carriage of toy poodle on RAAF aircraft. Approval is granted, on condition that the batteries are removed from the toy poodle prior to flight ...."

~ ~ ~ ~

Before the introduction of common place air travel, the Australian outback mining community of Broken Hill was serviced from Sydney by just two or three DC-3 flights per week and when the Douglas came to town it was still a big event. Kids would pedal out to greet it and it was not uncommon for their parents to often find an excuse to be at the airport around the same time.

Anyway, one house also played host to a wily pet parrot that obviously had better hearing than the average homosapien. The bird, obviously amused by the mayhem that took place whenever the DC-3 overflew our house and everybody rushed out to wave, thought that it could better this.

Hence, within a short period the bird learnt to identify the distant drone of the Pratt & Whitneys and begin bellowing at full strength "Here comes the plane, here comes the plane!" These were the same words the humanoids would yell whenever they heard it coming. This gave my family enough time to get into the yard before the DC-3 arrived rather than forever watching it depart.

Imagine something as simple and joyous as that happening today!

Douglas DC-3

~ ~ ~ ~

Near the end of the 1988 Australian Bicentennial Airshow, in the Richmond Back Bar, two RAAF aircrew from the same Pilot's Course ran into each other after several years absence.

One had achieved his ambition of becoming a 'knucklehead' on Mirages, while the other moved into the world of 'trashies' – a C-130 Hercules driver.

As old times were relived, it became obvious that each was fiercely proud of his particular aircraft. Over many drinks, they began extolling the relative virtues of their beloved charges – speed versus range, ordnance versus load capacity, manoeuvrability versus comfort.

This continued for several hours until, as closing time approached, the Herc pilot stood and prepared to leave.

"Listen mate," he said, "I'm really glad that your happy with the Miracle, and I'm sure it's a fine aircraft. But me, I'd rather have four screws than a blowjob any day!"

~ ~ ~ ~

Back in the 1950s, aircrew in the USAF flying C-97 Stratofreighter, the military version of the Boeing Stratocruiser four engined airliner, often got bored with carting US Army personnel all over the US and Canada to various exercises.

The routine flying got very monotonous, so one crew member of a Stratofreighter thought he had better bring some humour to the flightdeck. The Stratofreighter had a raised flightdeck which meant the crew were seated above and out of sight of their unwary passengers behind and below them.

On this particular flight, the bored soldiers in the back noticed a crewmember climbing down from the flightdeck in a rather unusual fashion. He had two long pieces of string in his hands, which originated from the cockpit. Once among the passengers, the crewmember walked backwards towards the rear of the aircraft, in a rather careful sort of fashion. The crewman, the copilot, soon found that he had run out of string, and so he asked a soldier if he would please hold the strings while he went to the rear of the aircraft to answer a call of nature, explaining that the soldier would be the person that was flying the aircraft for a few minutes and to hold the two strings very carefully.

The soldier wasn't impressed, saying he should try that gag on the Marines. The copilot somehow managed to keep a straight face, and said in all seriousness that he would demonstrate to him and the rest of the amused passengers that this indeed was no joke and that they were a little short handed on the flightdeck tonight.

A gentle pull on the left string, and the aircraft banked over nicely to the left. The passengers' amused looks turned to gapes of horror as the big aircraft was brought back to the straight and level position. A pull to the right, and the aircraft banked over rather steeply to the right. With the aircraft now straight and level, the copilot's final words before departing for the loo was, "It's all up to you son! Keep the tension on the strings and I'll be back in a few minutes!"

The soldier was left standing like a statue, with other soldiers looking on a little too grim faced to say the least.

C-97 Stratofreighter

Not quite to where he was heading, the copilot turned and yelled at the unsuspecting soldier some words of encouragement. The soldier, wanting to make sure he heard him correctly, did a half turn on the spot, which caused the strings to send the aircraft into a steep bank. The soldier fought to regain control of the aircraft while the other passengers screamed various forms of encouragement. The soldier, after his momentary loss of control, didn't know how much height he'd lost, whether he was still on the right course or if he should grab a parachute and make a safer trip back to terra firma. Once the aircraft was straight and level again, the crewman continued his walk to the rear of the aircraft.

The copilot finally made it to a private section of the aircraft, and burst into laughter. A few minutes later, after the crewmember had composed himself, he returned and carefully took the two strings, off the now very pale faced soldier. With the same careful manner that he had made his way among the passengers, the crewmember now made his way back to the cockpit. Once inside the cockpit the whole flightdeck erupted into raucous laughter as they removed the ends of the string from the epaulets of the captain.

Need we suggest it, but it was a very quiet group of soldiers for the rest of the flight.

~ ~ ~ ~

Back in the early fifties people mostly still travelled by DC-3s and DC-4s and aircraft sometimes had 'plug trouble'.

On the day of our particular tale a DC-4 had been chartered by members of Melbourne's horse racing fraternity to attend the annual Launceston Cup meeting. At their mid morning departure time the passengers took their seats only to learn that the aircraft had developed a mechanical snag. The passengers returned to the lounge and were told to expect a delay of about one hour and a message was sent to Launceston advising of the expected departure time.

With the hour gone by with no more news, a Race Club representative enquired at the airline office in Launceston as to how things were progressing at Essendon, as some jockeys engaged for the Cup were among the delayed passengers. An airline employee drafted a teletype to Essendon which read: "Launceston race official requests information with regard to charter flight. When this was reduced to telegraphese

Douglas DC-4

it became: LT RACES OFF REQUESTS INFO RE CHARTER". (In those days, ports had two letter abbreviations).

As it happened, the passengers were just boarding when the teletype arrived, and as there had been bad weather in Tasmania the apparent cancellation caused no surprise. Launceston was duly telephoned and told that the flight would not operate and passengers were being dispersed.

Fortunately, when the error was discovered, a public address message rounded up the passengers before they strayed too far and again they went aboard, this time making it to Launceston. Apart from a delayed start for the Cup, the mix-up ended on a happy note!

~ ~ ~ ~

A little old lady travelling on her first A320 flight asked the flight attendant the way to the ladies room.

The FA was busy with trays, so she pointed to the front of the aircraft, adding with a smile that she couldn't miss it.

But the old dear followed directions too literally. She walked all the way to the front, opened the door to the cockpit, and looked in at the crew at the controls. The two pilots swung their heads around in surprise as the old lady quickly slammed the door shut.

Confused, she returned to the galley and complained gently "There are two men in the toilet and they're watching television!"

Airbus A320 cockpit

Airbus

~ ~ ~ ~

Have you read Cdr John Nichol's and Barrett Tillman's excellent book on USN air ops over North Vietnam, *On Yankee Station*?

This book is excellent reading and includes many real world examples of how not to fight a war. That aside, there was one particular item in the chapter on aircrew rescues that was simply too good to not recall here. I quote:

"In March 1967 an RA-5C Vigilante from the USS *Kittyhawk* was flying a recce sortie at 350ft, just off the beach. Making 450kt (835km/h), the plane was badly hit from shore based AAA and went down in the water. The pilot was killed but the backseater ejected.

"The young lieutenant landed on the beach and had waded into the water before being surrounded by numerous hostile locals. He carried the standard issue .38 revolver in a bandoleer shoulder holster, but for added safety he left the first two chambers unloaded, using only four rounds in the gun. He also had a .22 automatic hidden under his flight suit. Thigh deep in the surf, surrounded by angry Vietnamese, the NFO knew that resistance was pointless. But he could see the duty SAR helo rapidly approaching from the east and was willing to try anything with the smell of rescue so close.

"One of the Vietnamese removed the .38 from its holster and pointed it at the flyer. Another local militiaman covered him with a rifle. The crowd dispersed when A-1 Skyraiders and F-8 Crusaders from the ResCAP began strafing up and down the beach, suppressing any anti aircraft fire that might impede the rescue.

"The airborne pilots could see the three men in the water and adjusted their passes accordingly. They could also see that the pilot was covered by two guns, but they couldn't fire at the Vietnamese without hitting the American. The most they could do was to provide distraction and harassment.

"That however was enough for the cool flyer in the water. When his potential captors glanced upward, taking in the low level flybys, he unzipped his flight suit, produced his hideout gun, and instantly chambered a round. With the two locals still distracted, he shot the rifleman in the head. The noise of the .22 round startled the other Vietnamese, who reflexively pulled the trigger of the .38. The hammer fell on an empty chamber, as the Viggie NFO knew it would. He shot the pistol holder, and in seconds was swimming madly out to sea. The helo swooped in and picked him up".

According to Nichols and Tillman that story made the rounds through Task Force 77 and back to Cubi (the Subic Bay HQ for the 6th Fleet air group) in little longer than it takes to relate the incident! Legend records that the level headed flier from "Heavy 13" was assured free drinks at the O Club as long as he was willing to tell his story!

~ ~ ~ ~

Relations between countries of differing cultures calls for a lot of understanding and diplomacy when it comes to dealing with each other's sensibilities.

In recent times Sweden has been keen to expand the sales prospects of its weaponry outside its traditional customer base, and so was highly keen to put on a good show when Pakistan asked the Swedes to give a presentation on the Gripen fighter. However the chances of the potent little fighter ever seeing Pakistani military service seem to have been dealt a telling blow, according to a report recently published in *Jane's Defence Weekly*.

When Pakistan's army chief, General Abdul Waheed, visited Sweden, he had the opportunity to inspect the Stockholm based mounted Royal Guard. So impressed with the Guard was the General that he decided to make a contribution of three prize Pakistani thoroughbred horses – Tez, Sardar and Sher Dil – as a goodwill gesture to his Swedish counterpart, General Ake Sagrén.

General Sagrén accepted General Waheed's generous gift despite being aware that Sweden's strict health regulations prohibit the importation of animals from Asia. The Swedish general was advised to accept the horses to avoid offending the Pakistanis and thus potentially jeopardise future military sales.

Unbeknownst to Tez, Sardar, Sher Dil and General Waheed, Sweden's restrictive rules meant that the three horses were promptly destroyed on their arrival in Scandinavia, despite the nature of the gift and cultural sensibilities involved! Army chief or not, General Sagrén was not able to influence Sweden's bureaucratic health officials, and the horses were doomed as soon as they left Pakistan, despite their prime condition.

Understandably, Pakistani Foreign Ministry officials were said to have expressed surprise at the horses' fate, saying that General Waheed would not have been offended if he knew that the horses' very lives were at stake. General Sagrén personally visited Pakistan to apologise for the horses' untimely fate and to try and explain the nature of his predicament.

While the Pakistanis reportedly gave him a little understanding, one can't help thinking that the chances of seeing Gripens in Pakistani colours are very, very small!

~ ~ ~ ~

Many years ago onboard a Constellation, in the days when navigators used to insert a sextant through the astrodome in the ceiling of the cockpit to take star positions for navigation, a standard trick was often played on new flight attendants.

The new recruit would be summoned to the cockpit while the navigator was taking star readings. Naturally she would enquire as to what the navigator was doing. Truthfully the crew explained that he was looking at the stars. She would naturally take the bait and ask if she could also have a look. Cleverly the crew would place their orders and agree for her to have a look on her return.

On her return the navigator would pretend to be looking at stars, and invite her to have a look. She would then complain that she could see nothing but pitch blackness. The navigator would explain that she was looking with the wrong eye. She would then change eyes and yet again see very little. The navigator would then help out by cleaning the lens and telling her to try again.

She would then clearly see the stars and eventually return to the cabin, and wonder why passengers would be giggling at her and not taking her seriously, until one of her fellow flight attendants took her to the nearest mirror to show her why. Horrified, she would suddenly realise why she could not see anything at first when she looked through the lens. The navigator had coloured in the lens with pencil, and she had been had – with two massive black eyes!

Constellation

~ ~ ~ ~

Spotted approaching Auckland's scenic international airport was a Qantas 767. Suddenly, there was a strong Australian accent delivering the callsign "Argentina 866".

AUCKLAND APPROACH: "Argentina 866, cleared to two-five zero-zero feet until you are established. Call the Tower 118 decimal 7" (followed by some very basic Spanish language with accompanying background laughter).

What followed showed that the Qantas 767 on charter to Aerolineas Argentinas had a flightdeck crew with a sense of humour.

ARGENTINA 866: (Qantas 767) "Look pal, I'm having enough trouble with handling the bloody callsign, let alone trying to speak Spanish as well!"

The regular Argentine service had an aircraft which had gone unserviceable in Auckland, stranding Argentina bound passengers in Sydney, where the flight had been scheduled to "turn around". Subsequently, a Qantas 767 had been chartered to allow these Sydney pax to join the aircraft in Auckland.

Boeing 767

Bill Lines

~ ~ ~ ~

Heard while monitoring Townsville Control frequency, which controls airspace in a 90nm (165km) radius around Townsville, was the following conversation between two Ansett 727s. VH-ANF was out of Cairns bound for Sydney, while VH-ANB was out of Brisbane for Cairns.

ATC: "Alpha November Foxtrot, request Townsville position report."

Then ... thirty seconds later ...

ATC: "Alpha November Foxtrot, request Townsville position report!"

VH-ANF: "Alpha November Foxtrot ... ahh ... can you clear something up for us. As we are on radar, do we still have to report over Townsville?!

ATC: "Alpha November Foxtrot, Townsville Control, affirmative, this is your last position report on radar, as we only have radar coverage to 90 DME south of Townsville."

VH-ANF: (now satisfied ...) "Okay, thanks, we were over Townsville at five-six, on climb to flight level three-three-zero, estimate Emerald at two-niner."

ATC: "Alpha November Foxtrot, thank you, report cruising."

VH-ANF: "Alpha November Foxtrot, roger!"

VH-ANB: "Don't argue ... just do it!"

VH-ANF: "Up yours!"

The last two comments were very much said in jest, as both pilots were obviously known to each other and the best of mates.

~ ~ ~ ~

VH-MVW, a Shorts 360 inbound to Gladstone in north Queensland called regional carrier, Sunstate Airlines' company office and the following conversation took place:

VH-MVW: "G'day Gladstone, we're about eight minutes out, we have 14 pax for you today, and we also will require some motion lotion!"

SUNSTATE GLADSTONE: "14 passengers, and what was the rest of it?"

VH-MVW: "Some firewater!"

SUNSTATE GLADSTONE: "Some what?"

VH-MVW: "Power kerosene!"

SUNSTATE GLADSTONE: "What ... err, why ... err"

VH-MVW: "Some bloody fuel for the flaming aeroplane!"

SUNSTATE GLADSTONE: "Ohh ... you want some gas ... I'll arrange it."

Shorts 360 VH-MVW

Bill Lines

~ ~ ~ ~

This story occurred during a very busy period of the 1992 Avalon Airshow, when many private and charter aircraft were arriving.

A Cessna 172 had just landed, and on a very short final behind this aircraft was a Seminole. As things got closer it was realised that Seminole was too close behind the just landed Cessna, so the controller gave him the next best option: overfly the Cessna coming to a stop on the runway, and land well ahead of him! The transmission went something like this:

ATC: "Alpha Bravo Charlie (the 172) ... you're clear to land, minimum delay on the runway thanks, there's a light twin right behind you."

VH-DEF: "Ahh ... Delta Echo Foxtrot, we're the Seminole on short short final, I frankly don't think we can make it down behind that 172!"

ATC: "Okay ... Delta Echo Foxtrot, an option. Can you overfly the Cessna and land ahead of him?"

VH-DEF: "Yes".

ATC: "So do it Sir!"

Not the usual requests you would hear, but it probably made good common sense considering Avalon's north-south runway is some three and a half kilometres long.

~ ~ ~ ~

A country pilot was approaching Adelaide, and was obviously not used to a controlled airspace environment. He also spoke in a very slow country drawl to a very busy ATC man.

COCKIE: "Adel-aide Con-trol, this is Char-lie Oscar Charlie, I'm forty five miles from Adelaide and I re-quest an air-ways clearance."

CONTROL: "Charlie Oscar Charlie, this is Adelaide Control, you are identified forty five miles from Adelaide. Your clearance: track your present position direct Edinburgh then Parafield, maintain 5000 and squawk code 4-4-1-4!"

COCKIE: (even slower now ...) :Ade-laide Con-trol, please re-peat slow-ly, I receive at the same rate as I speak!"

~ ~ ~ ~

At the Tom Reilley Warbirds Airshow a beautiful formation of a P-51 Mustang, T-33 Shooting Star jet trainer and a MiG-15 taking up the rear position took to the air one morning with the following conversation taking place.

T-33 (to MiG): "It's your formation, so it's your call!"

MiG-15: "Okay ... I call GUNS COCKED and LOADED!"

~ ~ ~ ~

A quickie from the Sydney airbands.

SYDNEY TWR: "Tango Bravo Golf, as you overflew it, was there much traffic on the Gladesville Bridge? (A large multi-lane bridge entering Sydney spanning 2-3km)

VH-TBG: "I don't even know where the Gladesville Bridge is !"

TOWER: "Never mind ... thank you anyway."

VH-TBG: "Err ... we just passed over a large highway and there wasn't much traffic on that."

TOWER: "That's the one ... thanks!"

The controller was a female. Shortly afterwards her shift ended, and a male voice took over the Tower frequency. The outgoing Aerodrome Controller was probably inching her car through the suburbs from the Airport ... towards ... you guessed it ... the Gladesville Bridge!

~ ~ ~ ~

This exchange was heard around the Sydney area (pre the third runway) between a heavily laden Qantas 747 and the Clearance service.

QANTAS 1: "Sydney Clearance Delivery, this is Qantas 1, stand 2, requesting clearance to Bangkok."

SYDNEY ACD: "Qantas 1, good afternoon, clearance is a West 1 departure, Flight Level three-seven-zero, and squawk code 2731."

QANTAS 1: "Roger, clearance West 1, Wollongong, FL 370, Squawk 2731."

SYDNEY ACD: "And Qantas 1, wind is zero two zero degrees at 15 knots. Do you want runway zero seven or three four for your departure?"

QANTAS 1: (After a lengthy pause) ... "Well, we could take zero seven (short runway – 2529 metres) but I forgot to put our rocket engines on!"

The crew was immediately allocated RW34. Wonder why?

~ ~ ~ ~

An article in a local newspaper concerning the Albion Park Airshow, south of Sydney, is a classic case of a journalist not familiar with aircraft getting out of their depth when trying to write about an airshow.

Various aircraft at the airshow were misnamed, as in the following list: Lockheed became Lockhead, Oshkosh Fire was labelled Oskkosh Five, Beechcraft as Beechcroft, Winjeel was referred to as Wingeel, a Stampe was written up as a Stomp, a Pitts Special as a Pitt, and finally an Extra 300 as a Exit 300!

~ ~ ~ ~

A scene set in northern Victoria, where two sport parachute drop zones (DZs) are located only a few kilometres apart.

Our hero (who we will call Ralph), is a parachuting instructor, and Ralph goes to the first drop zone to do some instructing. That evening, he journeys to the second DZ to stay overnight.

As he has some further instructing to do at the first drop zone, he gets a mate to give him a lift in a car back there, and with the jump complete, Ralph bums a ride in the Nomad drop aircraft over to the second DZ where he left his car.

The spotting is a bit rough, and Ralph our hero comes down under canopy to wind up upon his butt a few paddocks short of the drop zone. Does this worry our man? No, of course not! Quick as a flash his jumpsuit is unzipped, and he whips out his mobile phone and requests a pick-up vehicle to come and retrieve him. Ah ... the marvels of modern parachuting paraphernalia!

GAF Nomad

~ ~ ~ ~

Some aviators are still coming to grips with the fact that some ground stations are using handheld transceivers that are capable of scanning numerous frequencies in the airband allocation.

An incident that set a couple of pilots back on their heels occurred some years ago at a remote inland location in South Australia.

Two helicopters were inbound for a refuelling stop, calling the refueller on the discrete frequency to request fuel on arrival. The response came back in the form of the pleasant tones of a female voice.

"Sounds a bit of alright" said one of the chopper pilots to the other, after switching back to their chatter frequency, 123.45 MHz, the "numbers" channel.

"Naw," came the reply, "if it's the same lady who was here last time I was through this area, she's fat and 40!"

The first helicopter duly landed, with the pilot alighting from his machine to be confronted by a well built woman, no longer quite in the prime of youth, hands on hips, with a threatening scowl on her face. "Well, do I fit the picture?" she demanded of him. A very red faced helicopter pilot was heard to mutter that it was his mate who had made the remark on the radio earlier.

Fortunately, the refueller was a good natured soul with a sense of humour to match, and they got their fuel and, suitably embarrassed and hopefully much wiser, departed.

~ ~ ~ ~

The scene is one of former Australian regional Eastwest's BAe 146s VH-EWL, climbing to FL 240 from Brisbane, talking with Brisbane Control.

BRISBANE CTL: "Echo Whiskey Lima, report leaving flight level 1-8-zero."

VH-EWL: "Echo Whiskey Lima, roger."

Some minutes later ...

BRISBANE CTL: "Echo Whiskey Lima, present level now?"

VH-EWL: "Flight level 1-7-6 climbing!"

Later ...

VH-EWL: "Brisbane Control, Echo Whiskey Lima is passing flight level 1-8-zero."

BRISBANE CTL: "Roger."

VH-EWL: "We'll get there eventually!"

BRISBANE CTL: "Echo Whiskey Lima, elephants and pigs spring to mind Sir, but at the moment I can't tell you why!"

Unfortunately, this is characteristic of the BAe 146's performance in hot Aussie territory. It climbs slowly.

~ ~ ~ ~

A birdstrike had just occurred on an RPT 737 service out of Townsville, with the aircraft concerned gulping up a sizeable brolga (a large crane) into one engine on its take-off roll. The "tennis team" made a note to keep a sharp lookout for birds and other wildlife, and inform the Tower accordingly.

The "tennis team" was carrying out a daily servicing on the hook cable near the flight strip at the northern hook cable site, when one of the "team" spotted a large brown bird just outside the strip. He radioed the Tower, advising them of the problem, expecting the FAC Safety Officer to arrive to do his thing chasing the big bird away.

TENNIS TEAM: "Ground, Tennis Team, we have vacated the northern hook cable site, and for your info there is a large brown bird just outside the flight strip at the northern hook site. I think it's a bustard."

GROUND: "Well do the right thing and chase the bustard away!"

~ ~ ~ ~

This excellent piece of advice was heard during a torrential rainstorm on the Essendon ATIS.

"Melbourne Terminal Area Severe Turbulence Advice not available due to lightning strike!"

~ ~ ~ ~

This story occurred just before midnight during the Australian pilots' dispute, back in 1989.

At the time of the story, the dispute had been decimating domestic airline operations for some weeks. The Melbourne Senior Tower Controller (STC) had called in sick, and due to the industrial trouble, had not been replaced in a questionable decision by CAA management.

Although there was actually another controller on duty in the Tower, the Tower couldn't officially be operated. So the CAA made another dubious decision and declared the aerodrome closed!

However, a flood had occurred in a mine in Western Australia, trapping a number of miners underground. Pumping gear was urgently needed to be transported from Melbourne to Perth. An Ansett 707 freighter, VH-HTC, was operating Ansett Flight 471 inbound to Melbourne, to pick up the pumping equipment. Due to the aerodrome being officially "closed", AN471 declared itself a "mercy flight", enabling it to break the rules and land at Melbourne anyway.

Ansett 471 was on a control frequency, 123.6 MHz, and requested the surface conditions at the field, which were then passed to it by the Sector Controller. Ansett 471 then wanted to know how accurate the information was:

ANSETT 491: "Is there a physical person in the Tower?"

UNKNOWN: "No, it's an Air Traffic Controller!"

Melbourne Control was not at all amused, and even tried to unsuccessfully identify the culprit.

~ ~ ~ ~

Some international airline operators were not enamoured of the Sydney simultaneous runway operations (Simops) that were once in use, according to comments heard on the airways at the time.

Lufthansa 796, a 747-400, was on final for RW16 when Simops were in progress. A smaller aircraft was turning final for RW07 and was advised by Sydney Tower that "traffic is landing on the crossing runway, holding short of Runway 16". The ATC instruction was duly acknowledged.

Lufthansa 796 was then told that: "Traffic landing on the crossing runway will hold short and you are cleared to land!".

The grunted acknowledgement from a rather surly sounding pilot on Lufthansa 796 was not at all good natured. On short final, Lufthansa 796 requested confirmation of the landing clearance:

TOWER: "Affirmative, clear to land, The Super King Air landing on Runway 07 will hold short of Runway 16."

LUFTHANSA: "How do you know?"

~ ~ ~ ~

A Hazelton Saab 340B, registered VH-CMH, was on approach to Sydney from the north. A nor' easterly breeze was blowing across the airfield, according to the ATIS, and a rare quiet interlude was in progress.

VH-CMH was only a short distance from the field and the pilots obviously wanted the most expeditious approach under the prevailing conditions. A series of radio exchanges conveyed the strong implication that Charlie Mike Hotel was hopeful of a northern approach, which suited their relative position to the northwest of the airport.

The following conversation then took place:

VH-CMH: "What's the downwind on one-six?"

TOWER: "Ten to fifteen knots."

VH-CMH: "We can still take one-six if that's okay?" (stated very keenly)

TOWER: "Well, it's not the wind that is the problem Sir, it's the 747 on finals at the other end that concerns us!"

VH-CMH: "Oh ... disregard ... yeah ... "

VH-CMH was immediately given clearance for a RW07 arrival from the west and made a truly spectacular left turn for what was the shortest short-short final approach that had been seen in recent times.

~ ~ ~ ~

This particular RAAF Macchi was being ferried to Perth for repaint and refurbishment prior to delivery to 2 Flying Training School (2FTS) at Pearce. The aircraft had been used by the Chief of Air Staff of the time, AM Ray Funnell. It was a particularly sporty looking machine, with outstanding white, red, and blue trim. It also had a distinctive radio callsign, "CHIEFTAIN ZERO 1".

Over the years, Perth Tower ATC personnel had seen many Macchis pass through Perth on the way to depot servicing. On this particular ferry day, the pilot had obviously had a long day, and he executed an approach and landing on Runway 24 at Perth. During the rollout, the following exchange took place:

PERTH TWR: "Nice paint scheme Alladin 247!"

ALLADIN 247: "Yeh ... not bad for an old girl."

PERTH TWR: "Guess when you're the Chief, you can have any paint scheme you want?"

ALLADIN 247: "You guessed it right my friend."

Aermacchi MB-326H

~ ~ ~ ~

One day, a British Airways 747, callsign "Speedbird 9", is monitored inbound to Sydney very early in the morning, the Air Traffic Controller getting nervous about the aircraft landing before curfew finished, at 5:00am local. The aircraft was for a Runway 34 arrival, approaching from out to sea.

APPROACH: "Speedbird niner, you have 30 miles to run, what is your expected landing time?"

SPEEDBIRD 9: "About two minutes to the hour Sir."

APPROACH: "Speedbird 9, your computer is worth much more than mine!"

SPEEDBIRD 9: "Actually Sir, I was using my fingers!"

APPROACH: "In that case Speedbird, we have the same computer!"

~ ~ ~ ~

This story centres around an airline situation in Atlanta, involving Delta and USAir (now US Airways). Most of the Delta pilots are "good ole boys" from the South. They are regarded by other airlines and themselves as "the ultimate professionals". When other crews indicate on ATC channels that they wish to communicate with Delta crews on 128.95 MHz (air-to-air), the Delta crews often as not ignore the request, considering it "unprofessional" to talk air-to-air in such a manner.

Not all that long ago at Atlanta's very busy Hartfield International Airport, a Delta 767 and a USAir 737 were taxying to the threshold for the active runway. Out of their window, the USAir flightdeck crew noticed the undercarriage nose gear locking pin still intact under the Delta jet's fuselage as it taxied! This was the radio chatter that followed:

USAIR 186: "Tower, can you have Delta 254 speak to USAir 186 on 128.95 please?"

ATL TWR: "Okay Sir ... err, Delta 254, go to 128.95 and talk to USAir there."

DELTA 254: "Negative, professional pilots don't need to use that frequency!"

USAIR 186: "I think he better talk to us Tower!"

DELTA 254: "I say again Tower ... we're not changing frequency!"

Not a thing more is said on the subject during the prolonged taxi, which took over 11 minutes. Delta 254 finally arrived at the head of the departure queue about 20 minutes later, as earlier aircraft departed.

Then ... some more radio chatter on Tower frequency from USAir, who was about to get even:

USAIR 198: "Tower, you might ask the Delta 254 if they think professional pilots also takeoff with the locking pin still intact in their nose gear!"

Needless to say, the horrified Delta crew realised their omission, and requested to taxi back to their parking gate. Fifty minutes later, Delta 254 got to actually depart from Atlanta Airport.

~ ~ ~ ~

The following aircraft callsigns have been changed to protect the guilty.

FIS: "Alpha Bravo Charlie, IFR traffic is Delta Echo Foxtrot, a Mitsubishi, departed Bundy (Bundaberg, in Queensland) three three, on climb to three thousand for Maryborough, estimating Maryborough at four four."

VH-ABC: "Copy the traffic."

FIS: "And Delta Echo Foxtrot, IFR traffic for you is Alpha Bravo Charlie, a Cessna 414, which is tracking direct Brisbane/Bundy, leaving Flight Level one two zero and on descent to Bundy at four six."

VH-DEF: "Copy traffic Alpha Bravo Charlie."
VH-DEF: "Alpha Bravo Charlie, this is Delta Echo Foxtrot, I think I have you sighted now, eighteen miles, maintaining three thousand, I should be in your twelve thirty to one o'clock low!"
VH-ABC: "Yeah, we're thirty one miles, passing nine thousand, we won't go below three till you confirm passing."
VH-DEF: "Delta Echo Foxtrot is now one two miles."
VH-ABC: "I'll correct that, we won't go below four thousand till you confirm passing."
VH-DEF: "Roger."
VH-DEF: "I'm centred about your one o'clock low."
VH-ABC: "Yeah, haven't gotcha! You got strobes on?"
VH-DEF: "No, don't have any strobes unfortunately. Now at twenty four miles."
VH-ABC: "Two seven now. Got two lights on have you?"
VH-DEF: "Yeah, that's it, the taxi lights."
VH-ABC: "Right, gotcha now, gosh ... I thought it was a car!"
VH-DEF: "Well we're a big car ... a very fast one!"
VH-ABC: "You seemed to be doin' a good pace. Okay, have a good time."
VH-DEF: "Okay mate, catch you later!"

Mitsubishi Mu-2

~ ~ ~ ~

In this story, set at a fairly large airport in the American midwest, an instructor and student are holding on the runway in their Cessna 172, waiting for departing traffic on the cross runway. Suddenly, a deer runs out of nearby bushland, stops in the middle of the runway, and just stands there looking at them.
TOWER: "Cessna Five four Golf, cleared for takeoff."
STUDENT: "What'll I do Tower? There's a deer in front of us."
TOWER: "What do you think you should do?"
STUDENT: "Maybe if I taxi up to him it'll scare him."
TOWER: "Good idea!"

The Cessna taxied up towards the deer, but the deer is macho, and firmly holds his spot on the runway.

TOWER: "I say again, Cessna 1-8-5-4-Golf, clear for takeoff Sir, runway three-six!" (now starting to lose patience)

STUDENT: "What should I do Tower?"

TOWER: "T-h-i-n-k!"

Student: "I'm t-r-y-i-n-g to think."

Then for no particular reason, the "flying" deer bolts for the woods from the active runway. Two seconds later ...

TOWER: "Cessna Five Four Golf, cleared for departure runway three-six. Caution, wake turbulence, departing deer!"

~ ~ ~ ~

This item recalls an IFR flight in Germany, with an American deep south "Good Ole Boy" flying a USAF C-130 Hercules in a very crowded instrument pattern for landing at Frankfurt's Rhein-Main. The conversation was recorded as such:

CONTROL: "Air Force 1733, you're on an eight mile final for Runway 27-RIGHT, and you have a UH-1 helicopter three miles ahead of you on final, reduce speed to 130 knots."

AF1733: "Roger Frankfurt, we're bringing this big bird back to one hundred and thirty knots fur ya!"

CONTROL: (a few seconds later) "Ahh ... Air Force 1733, helicopter traffic at 90 knots now, one and a half miles ahead of you, reduce speed further now Sir to 110 knots."

AF1733: "Ah, Air Force three-three, and we're reining this big bird back again to 110 knots now."

CONTROL: "Air Force 1733, you are three miles from touchdown, helicopter traffic now only a half mile ahead of you, reduce speed now to 90 knots!"

AF1733: (by now he's quite ticked off) "Now look Sir, have you any idea of the stall speed of this here C-130?"

CONTROL: "No I haven't but if you don't know, try asking your First Officer, he can probably tell you!"

C-130 Hercules

~ ~ ~ ~

An airline aircraft came up on an Air Traffic Control frequency with some very non standard information. The crewmember using the microphone was the Captain, who introduced himself enthusiastically to his passengers over the cabin address system, and then proceeded to tell them all about the aircraft and the passenger facilities on board, the meals, movies, and everything else he could squeeze in.

The frequency's Air Traffic Controller then chipped in quickly, just after the hapless Skipper realised his awesome mistake and unkeyed the mike.

ATC: "Oh ... well done Captain Edwards!"

Apparently the good Captain thought he had said more than enough, and decided not to reply to the stinging remark.

~ ~ ~ ~

HMAS *Melbourne* was about to conduct some flying operations just off the coast of Botany Bay, south of Sydney Harbour. The warship wanted to communicate directly with Sydney Tower and they did.

MELBOURNE: "Sydney Tower this is Warship Melbourne."

SYDNEY TWR: "Yes Your Warship!"

Quick? Just typical of the lightning fast ATC minds in the Tower and Area Approach Control Centre!

And another very interesting item relating to the warship *Melbourne* surfaced recently, and it all revolves around a First Class seat from a Qantas Boeing 747. Intriguing? Well, the bridge of the latest HMAS *Melbourne* (an FFG guided missile frigate) now has an added air of luxury, thanks to the introduction of a Qantas First Class seat, with the Ship's Captain the main beneficiary! The seat, surplus to Qantas requirements, was earlier shipped to Port Melbourne to be fitted to the newly commissioned warship's bridge. It carries on something of a tradition, started by Qantas when it donated a First Class seat so long ago to the previous warship of the same name.

The Captain told his Qantas benefactors that he is on call 24 hours a day, and the donated First Class seat offers him "the best possible comfort during long stints on the bridge". Qantas has also supplied seats to HMAS *Success* and HMAS *Westralia*!

HMAS Melbourne

RAN

~ ~ ~ ~

"Alpha, Alpha, Alpha" (dummy callsign), a Cessna 402, was inbound to Sydney, and after checking his DME with Sydney Approach on 124.4, the exchange went something like this:

VH-AAA: "Can you give us an altitude readout?"

APPROACH: "What's your indication?"

VH-AAA: "Seven thousand."

APPROACH: "Ours shows seven thousand one hundred feet, which is within tolerance."

VH-AAA: "Like everything else on this plane!"

APPROACH: "Yeah, our equipment is the same."

Later on in the descent, the crew of VH-AAA indicated they had switched over to a backup transponder, however the Approach Controller indicated it was not being received. Approach later sought verification that VH-AAA was in fact descending, as his readout showed the aircraft had climbed to 9000 feet!

VH-AAA then asked for a radio check, and when informed it was "readability 5", the crew member had a final say:

"Ah well Approach, at least something works!"

~ ~ ~ ~

Scene: Perth HF communications centre.

USAF Starlifter: "Manila, Manila, this is AMC 3411."

This continued for several minutes, with Manila maintaining deathly silence. Suddenly, Perth decided to chip in and complete the relay.

PERTH: "AMC 3411, this is Perth, go ahead and I'll relay."

AMC 3411: "Ahh, station calling AMC 3411, say again your callsign."

PERTH: "AMC 3411, this is Perth, Perth – Western Australia."

AMC 3411: "Oh yeh ... where's that?"

PERTH: "Perth – Western Australia, it used to be the home of the America's Cup!"

AMC 3411: "Oh right ... THAT Perth!"

Lockheed Starlifter

Doug Mackay

~ ~ ~ ~

An oldie but a goodie! The Aer Lingus captain who when asked about his height and position responded ... "I'm five foot six and sitting up the front!"

~ ~ ~ ~

And now a Mother's Day tale from the airwaves around Sydney. An RPT jet, flown by a male was on final for RW16, and calls Sydney Tower on 120.5 which is responded to by a female ATC.

AIRCRAFT: "Ah, Sydney Tower, Juliet Echo Tango on final RW16."

TOWER: "Juliet Echo Tango, continue approach."

VH-JET: " ... And have a happy Mother's Day!"

TOWER: (peeved) "Juliet Echo Tango, for your information Sir, I'm an Air Traffic Control Officer, N-O-T a mother!

~ ~ ~ ~

This story concerns a Citation executive jet crew conversing with Sydney control. The conversation went like this:

CONTROL: "Juliet Echo Tango, are you GNS or LNS equipped?"

EXEC JET: "We're flight management system equipped!"

CONTROL: "Huh?"

EXEC JET: "That's GNS equipped, INS equipped, DME equipped ... should I go on?"

CONTROL: "No, thanks very much!"

(OTHER UNKNOWN VOICE): "Yeh ... but you can still get lost Laurie!"

~ ~ ~ ~

Overheard at London Heathrow:

Scene: RW28, one very hot summer evening. Departures from RW28 make a dogleg turn to the south so as to avoid Windsor Castle and thus making a noise there, which would ... perish the thought ... disturb the royals!

A PanAm 747, getting airborne from RW28, suffered an engine failure on rotation.

PANAM FIRST OFFICER: "Ahh ... tower ... Clipper 6, ah ... we're gonna continue straight ahead runway heading and dump some fuel!"

HEATHROW TWR: "Are you aware, Sir, that to continue straight ahead on runway heading will take you overhead Windsor Castle? Her Majesty is currently in residence!"

CLIPPER 6: (Quick as a flash ...) "Ask Her Majesty if she just wants the gas ... or the aeroplane and the gas!"

Boeing 747

~ ~ ~ ~

Overheard over Europe ...
Swissair 840: "Swissair 840, request flight level 330."
ATC: "Swissair 840, unable to give you FL330 due to noise abatement."
Swissair 840: "But we're now at flight level 290, how can we make a noise at FL330?"
ATC: "Swissair 840, we have traffic crossing you right to left at flight level 310. So if you climb now ... you will collide with that traffic and believe me, that situation *will* create a lot of noise!"

~ ~ ~ ~

Runway 13R at the Boeing Field in Seattle points directly at the 4395m (14,410ft) high Mt Rainer. The famous Mount Rainier is 65km distant but, because of its height, it looks more like 20 miles away. One particular day, a Learjet departed BFI via normal clearance, which was: maintain runway heading, expect vectors, maintain 11,000ft. The Lear did just that, but as you would expect passing 10,000ft, the pilot asked for higher, but was told to maintain assigned altitude.

The pilot was becoming very anxious indeed as he approached the mountain. After being denied his second request for higher altitude, he enquired of the controller the reason for the delay and was told 'noise abatement'. After an exchange of expletives, the controller calmly informed him that "there will be a hell of a noise if you hit the aircraft directly above you!"

~ ~ ~ ~

The Australian Department of Defence was experimenting with a radar controlled naval gun during the sixties. As the target towing aircraft flew over the range, a few rounds were fired up at the drogue, but as the radar was picking up a stronger return from the tow aircraft, the firing was slowly directed up the towline towards the tow aircraft.

Suddenly the tow aircraft directed this comment to radar control.
A/C: "Hey down there, I'm PULLING this mother ... not PUSHING it, okay?"

~ ~ ~ ~

An incident occurred one Sunday morning when Melbourne International was closed to all arriving aircraft due to fog. As a result, three jetliners were holding at Bolinda, north of the airport. Two were domestics, a 737 and a 727, and the other a 747 belonging to an Australian international airline.

The 727 crew advised that they had called Essendon Tower for conditions there and had been advised that they could make an approach to Essendon and land there if they wished. They decided to go and requested permission to track to Essendon, which was granted. The 737 crew was then asked by Melbourne control if they also would like to land at Essendon. The 737 crewmembers deliberated, but eventually decided to hold at Bolinda a while longer.

Then a voice broke in (obviously the 747 also holding) and said ... "Ah, Melbourne ... ah Australian 27 Papa, if we changed our name to something oriental, could we land at Essendon also?" It was a definite dig at an overseas airline whose crew recently mistook Essendon for Melbourne International and tried to line their 747 up for final approach! The final word on the situation came from the Melbourne Controller ... "Not nice!"

~ ~ ~ ~

A domestic DC-9 on descent into Melbourne some years ago, when Essendon was still being used as Tullamarine was not operational at that time.

The weather rapidly deteriorated on this occasion and the approach controller suddenly had his hands full with a lot of aircraft requiring instrument approaches. The initial navaid for the Essendon ILS is the Plenty NDB, so the controller gave the DC-9 a radar vector to point the aircraft in that general direction, hoping that they would make their own intercept of the ILS, so that he could quickly hand them off to the Tower and as such reduce the number of aircraft under his control.

The controller directed a question to the DC-9: "Tango Juliet tee-jet, are you getting Plenty?" The copilot was about to reply when the Captain shook his head and said to him ... "I've waited years for this!" He then transmitted the following to the stunned controller: "Our sex life is really none of your business Sir, however, in regard to other matters, we have intercepted the localiser if you would like us to call the Tower!"

~ ~ ~ ~

A British international airline, operating a Boeing 747 on approach into Sydney. The approach controller is trying very, very hard to speak clearly but is making a mess of it and the other voice is a very pukka British Captain.

ATC: "Silverbird one three, descend-ah to fife thousand."

A/C: "Silverbird 13, confirm descend to two-five thousand."

ATC: "I say again ... descend-ah to fife thousand."

A/C (starting to enjoy himself): "Silverbird descend to flight level two five zero, Silverbird 13."

ATC (getting frustrated now): "Silverbird one three, I say again, descend-ah to fife thousand."

A/C (now putting on his best 'humouring-lunatics voice'): "Ah yes, got it now ... descend to five thousand, confirm."

ATC (still hasn't got the message): "Silverbird one three, that is correct, descend-ah to fife thousand."

A/C: "Roger ... Silverbird thirteen, five thousand ... it's just that we'd have to go up again if you wanted two-five thousand!"

From ATC ... stony silence.

~ ~ ~ ~

From Canberra Airport comes the story of the day a giant USAF C-5A Galaxy arrived at Fairbairn to transport RAAF personnel, helicopters and support equipment for Australia's peacekeeping force in the Sinai.

Such an aircraft hadn't graced the runways at Canberra for a decade or more, so it attracted a lot of local interest. A road passing beneath the approach path to RW35 was lined with cars and budding photographers.

After a smooth, uneventful approach and landing, the road traffic quickly dispersed, just as a Department of Aviation Fokker F28 jet made a missed approach. To greet the Australian personnel, a welcoming party had apparently been formed consisting of persons in desert attire, complete with camels. Imagine the aerodrome controller's dismay when two of the aforementioned dromedaries were seen striding clumsily across the RAAF apron, seemingly out of control and bound for his active runway! Enter onto the scene the F28 jet, which was approaching short final. The resultant go round instruction allegedly included information not yet etched in aviation procedure, such as ... "runway occupied, two camels crossing right to left!"

The humpers were last seen literally heading west!

~ ~ ~ ~

This story concerns a young Australian flight attendant with Eastern Airlines in the US and a flight out of Boston bound for Miami where he decided that it was time to change the regular flight announcements that day.

They were about one minute out of Boston, on climb to the cruising altitude of flight level 330 and the young Aussie was that day on flight announcement duty at the microphone. He put a message to passengers over the PA which the flightdeck monitored, as usual, and which left the flightcrew with massive grins on their faces. "Ladies and gentlemen, just an Eastern advisory that this evening in Coach Class, rows 28 to 34, there will be no smoking. If you're seated in rows 10 to 14, there will be no singing and in rows 14 to 18, no dancing will be permitted. In rows 28 to 34 and 10 to 14, cigarette smoking only is acceptable, but no pipe or cigar smoking will be permitted. *Anyone* who contravenes these Eastern Air regulations will, I assure you, be asked to step outside the aeroplane!"

Later on, towards the flight's end, on descent to Miami, the Aussie wit again had the microphone in hand: "I'd like to remind all passengers to please check your overhead lockers and seat pockets in front of you for any carry on luggage, mothers in laws, or babies, that may have inadvertently been stowed there at the beginning of the journey. Thank you! I hope you enjoy your stay in Miami."

The last effort 'laid 'em in the aisles'.

~ ~ ~ ~

Australian Airlines Airbus VH-TAD, speaking to Sydney Control.

AC: "Sydney Control, good afternoon, Tango Alpha Delta, cruising flight level 350."

SYDNEY CONTROL: "Tango Alpha Delta, Sydney Control, good afternoon, identified one zero miles west of Parkes. Radar indicates you are about three miles to the left of track."

AC: "Tango Alpha Delta."

CONTROL: "Just a TAD to the left you might say!"

(Silence ...)

CONTROL: "Aww ... sorry about that."

AC: "We'll forgive you ... this time!"

Airbus Industrie A300 VH-TAD

Rob Finlayson

~ ~ ~ ~

Monitored one Christmas Eve was this:
CATHAY 101: "Darwin, Cathay 101."
DARWIN: "Cathay 101, Darwin, go ahead."
CX101: "Cathay 101, er, roger, er ... do you er ... have any traffic at our level, flight level 350?"
(Stunned silence from Darwin ...)
DARWIN: "Cathay 101, Darwin, negative."
CX101: "Ah, Darwin, Cathay 101, we've just passed very close to an unusual type of aircraft. It was in red livery and had a pair of skids for undercarriage. The Captain was dressed in a red flying suit and seemed to have a very long white beard. He actually called us on the VHF, but all he said was 'Ho Ho Ho! Merry Christmas Darwin!'

~ ~ ~ ~

Dateline: A busy Heathrow Airport and the 'star' of our story is a heavy jet airliner of US registration, requesting taxi clearance.
TOWER: "Yankee-two-one, clear to taxi, follow the 737 onto the outer taxiway for runway ten right."
A/C: "Roger, we'll follow the tin mouse for ten right!"
737 (A European operator): "Vee are not a tin mouse, vee are an American made Boeing 737!" (The crew are obviously a bit miffed!)
First A/C: "Er ... Roger, we'll follow the American made Boeing 737 tin mouse."
(Relative calm follows, with only occasional sniggering while the two aircraft taxi nose-to-tail for the runway. However, outright laughter ensues as the Tower has the last word ...)
TOWER: "Germanic 737 ... Squeak 3-5-0-4, clear for takeoff!"

~ ~ ~ ~

This story occurred a few years ago in a travel agency somewhere in the US.
There was a woman who wanted to fly to Europe for a holiday, but she was very worried about getting on a plane that had a bomb on board. No matter how much the travel agent tried to assure her that there was less than a one in a million chance of this happening, the woman had almost talked herself out of the holiday.
Just as the agent was starting to worry about losing the sale, his secretary piped up with the advice that the lady should carry a bomb on board herself, for the odds of getting on a plane with two bombs on board was virtually non-existent.
We never actually found out whether this advice convinced the woman to take her holiday or not.

~ ~ ~ ~

This story tells of media attention paid to the official opening of the Gateway Bridge over the Brisbane River. Various TV station helicopters were in the area for the opening and the following air/ground conversations were monitored between an overworked but totally professional controller and the helicopter pilots:
HELO PILOT: "And many thanks for your help. I bet you could use a beer now?"
CONTROLLER (quick to reply): "About two cartons!"

Later, a different helicopter pilot and the same controller ...
HELO PILOT: "You'll be glad when the day's over!"
CONTROLLER: "Nope ... not really, I'm working tomorrow morning and then I'm back on again tomorrow night."
HELO PILOT: "You must have done something terribly wrong!"

~ ~ ~ ~

Dateline: Bahrain Centre airspace and it's a typically busy night on Airway 'RED-19' from Bahrain westbound to Europe. With Iran airspace closed due to the ongoing war there is a large restricted area between FL260 and FL330 overlying the airway over northern Saudi Arabia, where AWACS aircraft continually patrol, monitoring the oilfields. A European operator is westbound from Bombay to Zurich. The following verbatim exchange occurs:
A/C (thick teutonic accent): "Bahrain Centre, good morning, European 999, flight level 310 and requesting higher."
CENTRE: "European 999, Bahrain Centre, good morning, flight level 350 is occupied, can you accept flight level 390?"
A/C: "I'm sorry Sir, we're too heavy for FL390!"
CENTRE: "Roger European 999, descend now to flight level 260."
A/C: "Christ Almighty, we have a fuel penalty already and now you want us to descend to FL260?"
UNKNOWN A/C: " ... and it's not such a good area around here for J C either!"

~ ~ ~ ~

One Sunday afternoon, a USAF C-141 Starlifter was overflying Brisbane inbound to Richmond when the following conversation was monitored on control frequency 123.0.
MAC: "Control, triple-zero-niner, bit quiet today huh?"
CONTROL: "Yes Sir."
MAC: "How's the weather down there?"
CONTROL: "Oh, not too bad. You're just passing over the top of Brisbane, just a few octas, very hot, about 31 degrees centigrade."
MAC: "Ah, it's not bad ... (few seconds pause) ... beach weather!"
CONTROL: "If I was there it would be."
MAC: "When we get there (Richmond) we're gonna grab a beer and head for the beach."
CONTROL: "It's a good move, but it's a fair drive from Richmond."
MAC: "Yeah."
CONTROL: "MAC zero-zero-zero-nine, is that the ... ah, military version of the Lockheed TriStar is it?"
MAC: "Negative sir, it's a four engine."
CONTROL: "Roger, the flightplan must have come out garbled, it has it down as a C-11, that's a C-141 is it?"
MAC: "That's affirm."
CONTROL: "MAC zero-niner, enjoy the beach, the beer and if you can make it ... the birds, contact Control 130.4. G'day."
MAC: "OK, thanks a lot, I've taken my vitamins ... ah 130.4 for the zero-niner heavy, have a happy day."
CONTROL: "You too."

~ ~ ~ ~

Something worth lifting from the Shirley Temple movie 'Bright Eyes'. Actor James Dunn was speaking to an old man who was seated in a wheelchair.

OLD MAN: "You're that aviation fellow, aren't you?"
DUNN: "Yes."
OLD MAN: "Well I don't like aeroplanes!"
DUNN: "Well I don't like wheelchairs!"
OLD MAN: "Stick to the aeroplane business and you'll wind up in one!"

~ ~ ~ ~

Here is a story about a 707 Captain with the sense of humour. Boeing 707 aircrew will remember the clatter and grinding that would emanate from the empennage of the great beast when on a particularly vigorous approach. This was largely caused by the screw jack system that operated the elevators and was often a cause of concern to novice flight attendants of the era.

The scene in question had our accomplished Captain bring in his 707 during a rather fierce storm at Hong Kong. With every nerve twitched and muscles flexed, manhandling the big jet between the rain squalls and wind gust over the populace of Hong Kong, the Captain is confronted with a junior hostess racing into the cockpit announcing – "Captain, there is a massive grinding noise coming from the back of the aeroplane!" Yells the Captain, without even breaking his concentration on combating the gyrations of the Boeing, "Well jolly well pull them apart!"

Boeing 707 on approach to Hong Kong

Qantas

~ ~ ~ ~

Overheard on a public address system at Williamtown where an airshow was in progress, some years ago.

ANNOUNCER: "Mr Phillips, your son says he will meet you behind the C-130 (pause) and the C-130, Mr Phillips, is the large grey and white aeroplane with a wing on top and four engines!"

~ ~ ~ ~

Scene: An ANZAC Iroquois helicopter serving with the Sinai peacekeeping force, returning to its base, at that time in the Egyptian Control Zone, located at El Gorah.

A/C: "El Gorah Tower, Mike Foxtrot Oscar 884, five miles south, inbound."

TOWER: "Mike Foxtrot Oscar 884, where you from?"

A/C: "El Gorah."

(Egyptian controller now pondering on a problem ...)

TOWER: "Mike Foxtrot Oscar 884 ... where you going?"

A/C: "El Gorah."

(Egyptian controller now thinking very hard)

TOWER: "Mike Foxtrot Oscar 884, you from El Gorah?"

A/C: "Yes."

(Egyptian controller now acting like lawyer in training ...)

TOWER: "Mike Foxtrot Oscar 884, you go to El Gorah?"

A/C: "Yes."

(Egyptian controller now really confused ...)

TOWER: "You wish land El Gorah?"

A/C: "Affirmative, Mike Foxtrot Oscar 884, we wish to land El Gorah!"

(Egyptian controller – now totally befuddled – gives up ...)

TOWER: "Mike Foxtrot Oscar 884, clear land!"

Defence PR

UH-1H Iroquois

~ ~ ~ ~

The following 'high level' conversation took place between 767 VH-RMG on a regular Sydney run, and F28 VH-EWD on a flight to Coolangatta. Both the aircraft were northbound. A line of thunderstorms was causing numerous aircraft diversions

VH-EWD (at FL290): "Hey, that Ansett high-flyer up ahead, how high are those storms?"

VH-RMG (at FL410): "Ah ... they're at 39,000 feet."

VH-EWD: "I think I'm sorry I asked."

A slight pause, then ...

UNKNOWN AIRCRAFT: "If God had wanted us to fly that high he'd have given us all Boeings!"

VH-RMG: "Sir! GOD MADE BOEINGS!"

~ ~ ~ ~

An Air New Zealand DC-8 cargo aircraft was ex-Melbourne enroute to Sydney with a load of racehorses.

TE 623: “Melbourne Control, New Zealand six two three on climb to flight level three seven zero.”

CONTROL: “New Zealand six two three, good evening, this is Melbourne Control, reporting leaving flight level one one zero, climb to three seven zero.”

TE 623: “New Zealand six two three, one one, for three seven.”

Shortly afterwards, the DC-8 is heard to call “leaving one one zero”, while the controller is kept busy with a Thai 747 crew, who seemed unfamiliar with the Melbourne FIR. With three minutes having passed, the controller enquires as to the DC-8's flight level and is told “New Zealand six two three, left level three one five!”

The controller then speaks in a surprised tone:

CONTROLLER: “Is that a normal DC-8?”

TE 623: “Yeah, well actually we are a little light tonight, with only one container below.”

CONTROL: “And a few extra horse power.”

The concluding scene must have been one fast DC-8 jet, with two pilots and a flight engineer sporting massive grins and the smart controller, once again back to his job of vectoring the Thai heavy.

~ ~ ~ ~

A Cessna Conquest and a British Airways 747 were the players in this scenario. The 747 was on an eight mile final for RW21 and Approach was enquiring if the Conquest had the 747 visual. The Conquest was being radar vectored for a RW24 arrival.

APPROACH: “Tango Foxtrot Golf, report sighting a 747 now on a five mile final runway two-one.”

(A few seconds pass ...)

VH-TFG: “Err ... Tango Foxtrot Golf, I don't have the 747 visual!”

APPROACH: “He's now on a three mile final.”

VH-TFG: “Negative ... no, I'm sorry, I don't have him visual.”

ATC came back with this fast one liner ... “Just look for his wallet!”

Cessna Conquest

~ ~ ~ ~

A pilot living at Kunanurra in Western Australia provides this rather amusing but also worrying look at an aviator who perhaps should not have been in the air:

AIRCRAFT: "Flightland, this is Delta Uniform Mike, taxying for local flight, copied ATIS, four POB."

FLIGHT SERVICE: "Delta Uniform Mike, go ahead your endurance."

AIRCRAFT: "Flightland, this is Delta Uniform Mike, taxying for local flight, copied ATIS, four POB."

FLIGHT SERVICE: "Delta Uniform Mike, copied all that, go ahead your endurance!"

AIRCRAFT: "What do you mean?"

FLIGHT SERVICE: "How much fuel do you have on board?"

AIRCRAFT: "Oh we've got two full tanks of fuel in this aeroplane!"

FLIGHT SERVICE: "How many minutes is that?"

AIRCRAFT (still unsure): "Oh ... standby."

This one must win the award for ineptness. I have heard several other examples of the same thing recently and it does make one wonder. The FIS place name and aircraft registration used in this story are fictitious.

~ ~ ~ ~

Just another day during the 9-5 period of the 1989 Australian domestic pilots' dispute.

QANTAS 2: (Inbound to KSA) "We'd like direct track to West Pymble."

CONTROL: "You might have trouble there, because of holding."

QANTAS 2: "Holding ... you have got to be kidding. You have heaps less traffic because the domestics aren't flying after five local and you try to tell me there will be holding!"

CONTROL (very polite): "I'm afraid so sir. Unfortunately, the way you've recounted is not exactly the case. Instead of airlines trying to cram 200 passengers on a 767, we're stashing 10 at a time on lighties and, as a result, traffic is increased tenfold. At the moment, it's murder for us here!"

QANTAS 2 (quite taken aback): "Err ... ah, well, could you check on that possible holding for us please and pass the expected delay?"

CONTROL: "Qantas 2, Sydney Control, roger, I'll advise you shortly."

(A few moments later ...)

CONTROL: "Ah, Qantas 2, Sydney Control I turned and asked the Flow (Approach Flow Co-ordinator) if you had holding and he just laughed at me! So I guess that means you have it and it's significant!"

QANTAS 2: (Indecipherable)

~ ~ ~ ~

UNITED 816: "Control, we're passing over what appears to be a very large meteorite crater. Do you know what it is?"

CONTROL: "I've noted the spot on radar Sir, I'll try and find out."

UNKNOWN A/C: "Have you checked your cargo door lately?"

~ ~ ~ ~

Just after a thunderstorm, now passing out to sea.

VH-TBG (727): "Tower, Tango Bravo Golf on long final for two-five, and we're really enjoying the pyrotechnics and light show you put on for our arrival!"

TOWER: "Aww thanks, we have spent some time working on 'em."

~ ~ ~ ~

On the particular night of our story, the crew of a USAF KC-135 is airborne and monitoring the exchanges on radio between a trainee air traffic controller and the crew of a high flying SR-71 Blackbird, callsign 'HABU 29'.

KC-135 FIRST OFFICER: "Captain, I don't think that controller's got the foggiest what a Blackbird is or what it can do!"

TRAINEE CONTROLLER: "HABU 29, understand you are requesting 80 thousand! Is that affirmative?"

SR-71 BLACKBIRD (HABU 29): "That's affirm Centre!"

TRAINEE CONTROLLER: "Well if you think you can make it, go ahead and try!"

SR-71 BLACKBIRD: "Roger, this is HABU 29 outa 90 thousand descending for 80 thousand!"

There was nothing but stunned silence from the hapless young Air Traffic Controller.

SR-71 Blackbird

~ ~ ~ ~

Two Ansett 737s are the stars of this next piece.

SYDNEY CONTROL: "Charlie Zulu Lima and Charlie Zulu Mike, this is Sydney Control. Information that Melbourne Control will not be accepting your radar contact at your present levels. Either of you should take Flight Level 350 or 390."

VH-CZL: "We'll take Flight Level 350."

SYDNEY CONTROL: "Roger, that's great, ah ... Charlie Zulu Mike, are you now able to accept Flight Level 390?"

VH-CZM: "Just a reminder ma'am, 737s only operate to Flight Level 370."

SYDNEY CONTROL: "Err ... roger ... standby!"

VH-CZL (directing comments to VH-CZM): "Good on you Dave, I thought for one ugly moment there that we were going to knock each other out of the air!"

VH-CZM: "If either of us were going to be knocked out of the air it would certainly be you mate!"

SYDNEY CONTROL (very angry voice): "Charlie Zulu Lima and Charlie Zulu Mike, cease all transmissions. Charlie Zulu Lima, climb to and report level at Flight Level 350, Charlie Zulu Mike, descent to and be at Flight Level 310 by time one-seven."

~ ~ ~ ~

TOWER: "Alpha Bravo Charlie, would you prefer left or right circuit?"

VH-ABC: "Affirmative Tower."

TOWER: "Alpha Bravo Charlie, which one would you prefer?"

VH-ABC: "I'd like a right circuit for a left base!"

~ ~ ~ ~

Another Sydney story. The aircraft involved was a DC-10, over Palm Beach for left base RW16. The Approach Controller was also controlling a queue of four small commuter aircraft bound for Sydney International.

US BASED DC-10: "Sydney Approach, Worldwide 215 heavy, we'd like a short approach, can you get those midgets out of the way?"

APPROACH: "Worldwide 215, maintain present heading, I'll give you a left turn in five miles."

DC-10: "Look ma'am, if you'd move those flying ants out of the way, I'd be on the ground in five minutes flat!"

APPROACH: "I say again, Worldwide 215, maintain present heading!"

DC-10: "Look lady! You may not know this, but we're a H-E-A-V-Y, so give us a break. Ever heard of a DC-10 ma'am?"

APPROACH (very, very cool and professional ...): "Worldwide 215 ... maintain present heading, you are now number six to land behind two 747 aircraft on approach to Pymble that are even heavier than you! Expect a left turn in another five miles!"

(Sweet revenge ... well done!)

~ ~ ~ ~

Our hero in this story is an experienced pilot with many hours logged up on twins. He had decided to refamiliarise himself with the Seneca, not having flown one for a while. He also just happened to have two cracked ribs as a result of a sporting accident that had happened a few days earlier.

As any Seneca flyer will attest, they are V-E-R-Y heavy aeroplanes to fly in the flare if they are not trimmed correctly on final. As our hero hasn't been in a Seneca in a long, long time, he forgets to trim the aircraft ... doesn't he. So ... during the flare, the stick gets heavier, heavier, and even H-E-A-V-I-E-R still! With a death like grip he grabs the column with both hands, simultaneously depressing his press-to-talk switch. With all the effort required, his cracked ribs are really hurting, and our hero lets out: "Stuff it, this thing is bloody heavy!"

To which our ever vigilant boys in the Tower respond with ... "Mike Charlie Kilo, say again all after ... *Stuff it, this thing is bloody heavy!*"

~ ~ ~ ~

Solo student pilot downwind.

AIRCRAFT: "November Echo Yankee, downwind touch-and-go."

TOWER: "November Echo Yankee, clear touch-and-go!"

AIRCRAFT: "Tower, is November Echo Yankee clear to land?"

TOWER: "No ... you requested touch-and-go!"

AIRCRAFT: "November Echo Yankee, going round!"

~ ~ ~ ~

NAVAJO CHIEFTAIN: "Tower, there are parrots eating near me on the runway!"

BANKSTOWN TWR (very laid back): "Really."

NAVAJO: "Maybe the Safety Officer could pay them some attention. There's a whole group of them. It looks like they're eating and having a chat!"

TOWER: "Oh ... no further explanations necessary. It's obviously not parrots, they're just management people from our Central Office having a meeting."

~ ~ ~ ~

This next one involves a RAAF Caribou callsign 'backduck' and the misinterpretation that followed.

BACKDUCK 215: "Perth, Backduck 215, circuit area Pearce this time, call again after landing."

APPROACH: "Roger, after landing, if no contact, try Apron Clearance Delivery 133.0, and traffic is Kilo Whiskey Whiskey, a Cessna 210, in your 12 o'clock, ten miles."

BACKDUCK 215: "Roger, copied traffic, and 133.0."

APPROACH: "Kilo Whiskey Whiskey, traffic is Backduck 215, your 12 o'clock, nine miles, make right turn 260 for separation."

VH-KWW: "Roger, 260 degrees ... what's a Black Duck?"

APPROACH: "Something that swims and goes quack!"

RAAF Richmond

DHC-4 Caribou

~ ~ ~ ~

Historical (err ... hysterical!) Section now. Circa 1958, with the star of the show a RAAF Canberra bomber, callsign VM-JBH, enroute Darwin to Amberley. The pilot is a USAF exchange officer, and the conversation between Isa Control and old 'Bravo Hotel' went like this.

VM-JBH: "Ah ... Isa Control, Victor Mike Juliet Bravo Hotel, an air force jet, Darwin for Amberley, one hundred DME Isa, estimate overhead Isa in one zero minutes, we're at flight level five zero zero."

CONTROL: "Bravo Hotel, Isa Control ... say again your flight level" (sounds confused!).

VM-JBH: "Flight level five zero zero, Isa!"

CONTROL: "Bravo Hotel, Isa ... you mean fifty thousand?"

VM-JBH: "You bet!"

CONTROL: "I bet you're a bit cool up there."

VH-JBH: "Waal ... if I wuz makin' a Tarm Collins, I sure as hell wouldn't need any ice. It's minus sixty four degrees outside for your info!"

~ ~ ~ ~

At the time a Fleet Replacement Pilot was being trained in the E-2C Hawkeye.

GCA: “Cloverdale 702, you have unknown traffic at twelve o’clock, three miles.”

E-2C (following a very diligent search): “Negative contact! Where’s the traffic now?”

GCA: “You’re clear of it now.”

(The approach was completed uneventfully, and while the Hawkeye was on the downwind leg for the next one ...)

GCA: “702, you have traffic on the nose again at two miles.”

E-2C (another diligent search ... again no joy ... and by now the instructor and his student are somewhat baffled): “Err ... no sighting of that traffic. Continue the calls please of where the traffic is!”

GCA: “702, do you see a flock of birds in front of you ... about a mile away?”

E-2C: “That’s affirmative, they’re very low. I didn’t know your radar could track seagulls.”

GCA: “Only when they’re squawking!”

Northrop Grumman E-2C Hawkeye

~ ~ ~ ~

A solitary Qantas 747 was making slow progress about three quarters of the distance to the threshold of RW34 at Sydney. The crew called “ready”, and after a short delay, the following conversation transpired.

TOWER: “Qantas 3, clear for takeoff, and ahh ... keep it rolling.”

QANTAS (singing the song Rawhide): ... “Rollin’ – Rollin’ Rollin’!”

TOWER (Qantas 3 just airborne): “Qantas 3 – Rawhide, contact departures on 123 decimal zero.”

~ ~ ~ ~

The next incident involves a Darwin based civil Bell 412 helicopter on SAR duties for the RAAF.

The captain of the helicopter was having trouble with one of the two helipilot buttons on the aircraft being stuck down.

In desperation, he pushed the intercom button to demonstrate his frustration to the other pilot, but due to some further finger trouble with the transmit-selector, the following call went out over Darwin Tower frequency: “It’s stiff and I can’t get it up!”

A very cool and professional sounding female Controller replied very quickly ... “I’m sorry, but I can’t help you with that problem!”

~ ~ ~ ~

A pilot was recently flying some circuits at Canberra to, as he puts it, "keep the dust off his CSU and RETRAC endorsements". He took a friend from work, who was at the time doing his Restricted Licence and who had been advised by his instructor to practice his radio procedures.

The workmate was doing fine, with little prompting, until they were cleared to touch and go, maintain runway heading. They were abeam Mount Ainslie by the time they received further instructions.

TOWER: "Charlie Sierra Hotel, make right circuit."

WORKMATE: "But there's a hill in the way!"

TOWER: "Miss the hill, then make right circuit!"

~ ~ ~ ~

Year: 1981. Dateline: Holsworthy Army Camp near Sydney.

Scenario: An airdrop into Holsworthy by RAAF Caribou from No38 Squadron at Richmond, assisting TLC Troop, 2nd Cavalry Regiment in the exercise.

OIC: "Captain Meathead."

The Caribou aircraft, laden with pallets containing full jerry cans heads towards Holsworthy getting clearance from ATC for the descent from 5000 feet. The pilot called the ground based soldiers with the following message: "Five-one, this is Enfield 306, do you read, over."

Captain Meathead responded to the call, then the aircraft requested latest weather at the dropzone, to which Captain Meathead replied with several minutes of totally meaningless information.

The aircraft crew then said "We're approaching you now", to which the good officer responded with ... "Yes, I can see you now!"

The aeroplane came closer, and Captain Meathead became alarmed at the erroneous track adopted by the Caribou crew ... way too far out to the left, away from the designated and painstakingly laid out DZ (drop zone).

The Captain relayed his fears over the frequency to Enfield 306, to which the pilot replied ... "Okay, then where are we in relation to you?"

Captain Meathead started a monologue of totally dumb, senseless and extremely confusing 'instructions', to which Enfield 306 retorted ... "Look, which way do you want us to go?"

The Army Captain ordered the aircraft "to turn more to the right", and was incredulous as the aircraft was piloted abeam of the DZ about half a kilometre, flying past several by now very amused troopers.

Captain Meathead then informed the pilot that "you are level with us and going past now". The Caribou was flown dead ahead, turned and disappeared behind a cluster of trees for a short period. The pilot advised "Going round, we'll come back in again!"

Sure enough, a few minutes later, the same aircraft appeared on the horizon and Captain Meathead again advised the Caribou skipper of an erroneous track, and started again ... more to the right ... more to the right!

The crew responded, but the Caribou stayed on the same flight path as before. Everything that happened the first time around happened again, much to Captain Meathead's chagrin.

On attempt number three, with Caribou and Meathead getting a trifle edgy, and the troopers having great trouble controlling their mirth, and making suggestions to the effect that maybe we have the wrong aeroplane Sir?

"Of course we haven't!" was the confident reply, followed by ... "Enfield 306, this is

Five-one, look I can see you clearly, you're doing exactly the same as before. I know it's you, I can see the red motif on the tailplane!"

To which the Caribou crew replied with, "Two things we better clear up now, firstly we are over the Blue Mountains at this time, and secondly we don't have a red motif on our tail, it's yellow!"

(Need I say or explain more, except to say the other Caribou was involved not in Dropzone activity but was carrying out a series of well executed touch-and-go manoeuvres at the nearby Luscombe Field!)

~ ~ ~ ~

Aerodrome Controller in Queensland to a departing aircraft we shall call Delta Uniform Mike.

ATC: "Delta Uniform Mike, what's your track to Kagaru?"

VH-DUM: "It's just on the railway line!"

ATC: "Which railway line is that?"

VH-DUM: "The one that runs through Kagaru!"

Yes ... our correspondent assures me it really did happen. How do these guys get a licence in the first place?

Another story, same contributor.

A Cessna turns incorrectly across the path of a Cherokee getting airborne from a parallel runway. After suitable chastisement from the Tower, the pilot of the offending aircraft replies ...

"Did we miss?"

Third story from the same anonymous guy ... A student pilot, obviously feeling guilty at missing a downwind call on the radio, lands, clears the runway, and then calls up on the SMC frequency ... "Ahh ... downwind ... full stop!"

~ ~ ~ ~

'I hit the wrong button' department.

Ansett's Charlie Zulu Juliet (VH-CZJ) was the star of this scenario copied on Melbourne/Launy Control 119.7.

VH-CZJ: "Good morning ladies and gentlemen, this is your Captain, my name is Ian ... , and First Officer Grant ... on the flightdeck this morning. We are presently passing through 10,500ft on climb to our cruising altitude of 33,000ft. Unfortunately, we are expecting a bit of rough air during the flight. As a result the 'fasten seat belt' sign will remain illuminated. If you feel that you have to move around the cabin at any time during the flight, please ensure your seat belt remains fastened!"

"If you have any enquiries about any matter during your flight, our flight attendants will be glad to help you. The weather in Melbourne this morning is a little chilly, thirteen degrees, and showers are expected."

CONTROL: "Thank you very much Ian!"

VH-CZJ: "Oh no ... I didn't ... did I Control?"

CONTROL: "I always wondered how I would move around the cabin with seatbelt still done up!" (This was said very sarcastically.)

VH-CZJ: "Oh God, did I really say that?" (extremely stunned)

CONTROL: "Perhaps better to tell me than the passengers!"

~ ~ ~ ~

This story involves an Australian 737 on company frequency talking with Transair about a 'small' problem. Well ... the people who caused the problem were small ... a group of young children on a group booking who must have left their point of departure with well meaning health conscious mothers!

AIRCRAFT: "Transair Mackay, Tango Juliet Foxtrot."

TRANSAIR: "Go ahead Foxtrot."

AIRCRAFT: "Ahh ... for catering we have some urgent servicing requirements, and the cleaners for that matter ... we'll need the aircraft completely cleaned out as the little buggers have stuffed orange peels down all the seat pockets!"

~ ~ ~ ~

This one involves VH-HYA, an Ansett A320 with an American pilot, ex Melbourne for Brisbane on the "rocket route", and VH-TJC or VH-TJE, the only other aircraft on frequency at the time.

HYA: "Ahh Melbourne Control, Hotel Yankee Alpha."

CONT: "Hotel Yankee Alpha, this is Sydney Control!"

HYA: "Ahh ... just checking if you're still there ... it's very quiet this morning Sir."

CONT: "Yes, there's not much traffic at the moment ... but then, that's the way we like it!"

HYA (sarcasm plus ...): "I can't get used to all the yakking in my ear here, unlike home of course!"

TJC or TJE: "Yak Yak Yak Yak Yak!"

Airbus Industrie A320

Bill Lines

~ ~ ~ ~

Right message to the wrong people department!

Apparently certain enterprising Qantas pilots have worked out a system of avoiding the "Please explain why you departed late!" note in their pigeon hole.

A very matter-of-fact Captain on board a Qantas 767 made this announcement over the aircraft passenger intercom, meant for crew/groundcrew intercom only:

"Ahh ... if we release the brakes and push the aircraft back six, the machine will record us as having departed on time!"

A few minutes later, realising his mistake, he apologised to the passengers, saying that "there were too many buttons to push, and that another aircraft wanted his parking bay," but apparently, every single passenger on board had seen through his explanation.

~ ~ ~ ~

This exchange occurred a few years ago in the oilfield country of far western Queensland where there was the usual banter between the "real men" (chopper pilots) and the "wimps" (flying fixed wings).

A Cessna 402 was having problems with the wheels not showing down prior to making an approach. There was lots of radio communication traffic from everyone aloft.

NORM (chopper pilot): (trying to be helpful) "Err, I'm twenty minutes away, and I can fly near you and look."

CESSNA PILOT (very agitated): "Look man ... I'm not waiting twenty minutes for you to inspect my plane!"

NORM (chopper pilot): "I don't want to inspect your plane, I just wanta watch you land it!"

~ ~ ~ ~

One day around noon in the country music capital of Tamworth, the only traffic was a Twin Comanche and an Eastwest BAe 146 doing circuit training. Arriving on the scene was an F/A-18 Hornet for a couple of circuits, before it headed on to Williamtown. The Comanche and BAe 146 had landed, with the 146 waiting on the apron for further circuits. The Hornet was on final for its lash touch and go, and the air/ground conversation went like this:

TOWER: "Despot one five, I have the Tower chock full of school children, and they would like to see how that F/A-18 you're flying goes straight up!"

F/A-18: "Tower, Despot one five ... sure ... can do!"

TOWER: "I must admit the controller would too!"

The Hornet was brought in for a classic touch and go and was then placed in vertical climb attitude, attaining Flight Level 150 in fifteen seconds, with the Tower controller turning his attention to the BAe 146.

TOWER: "Echo Whiskey India, Tower ... ok Sir ... your turn now!"

BAe 146: "Tower, Echo Whiskey India, I don't think I can match that thank you. We'll pass on that climb!"

~ ~ ~ ~

Remembering the great hoo-haa caused by that now famous television interview where the then Australian Prime Minister was caught out not wearing a seat belt in his limo, this next story is appropriate.

Helicopter VH-NWD, owned by the Roads and Traffic Authority of NSW, was being cleared to Bankstown Airport through controlled airspace. The pilot announced that "I am picking up the Prime Minister's party, who have been meeting the people as an aftermath of the Keating leadership challenge".

On return to Sydney Airport, the helicopter was cleared to land by the controller at the Qantas executive area. Then a voice broke in on the Tower frequency 120.5.

ANON: "You'd better make sure all their seat belts are done up!"

VH-NWD: "Hang on ... I'll just check" (and ... after checking ...)

VH-NWD: "He said he forgot ... and he's sorry!"

Nice to see the PM hadn't lost his sense of humour. .

~ ~ ~ ~

Overheard at Townsville was the following conversation involving the Tower, a Piper Warrior, a Twin Otter, and five RAAF F/A-18 Hornet jets.

The Warrior (VH-TEE) called ready for takeoff from RW01, followed closely by Australian Regional Twin Otter VH-TGH, who was ready at the centre taxiway. Both aircraft are instructed to hold position, due to Hipshot 5, the formation of F/A-18s, about to depart.

VH-TEE: "Townsville Tower, Tango Echo Echo ready!"

TOWER: "Tango Echo Echo hold, a short delay."

VH-TGH: "Tower, Tango Golf Hotel ready."

TOWER: "Tango Golf Hotel, hold position, short delay due military aircraft departing."

HIPSHOT 5: "Tower, Hipshot 5 ready."

TOWER: "Hipshot 5, clear for takeoff, make left turn."

Five RAAF F/A-18 Hornets then make a very spectacular formation departure.

Then ...

TOWER: "Tango Golf Hotel, clear for takeoff, make left turn."

VH-TGH: "Tango Golf Hotel roger, watch this!" (the aircraft then making a very normal departure)

TOWER: "Ooh ... very nice Sir!"

VH-TEE: "Yes ... the sheer speed and power involved was just awesome!"

VH-TGH: "We're just handing out the neckbraces now for the passengers who received whiplash from the sustained G-forces!"

Defence PR

McDonnell Douglas F/A-18 Hornets

~ ~ ~ ~

It's not often the Police come straight out and admit admiration for some spectacular criminal act, but the $1.3m midair heist carried out on an Air Inter (European) flight was described by investigating policemen in Paris as "brilliant".

A clever thief hid in a trunk inside the secured hold of the Air Inter plane in a successful bid to steal more than 5.7m francs, or about $A1.3m. The money was considered as safe as it could ever be, more than 30,000ft above the ground, in a pre-searched hold, with security guards on the plane. The cash had been collected from Corsican banks and post offices, and simply 'vanished' during the flight from Corsica to Paris. Even the Brinks Armoured Car Company, which fights a constant battle with the tough Paris underworld, was taken aback.

It was done this way: the thief eased the trunk open during the flight, clambered out, found the sealed money bag, broke the seal, moved the money into the trunk, stuffed the money bag with newspapers and cardboard which also had been carried in his trunk hiding place, and then crawled back into the trunk. The bizarre episode continued when the aircraft touched down in Paris, with guards from Brinks receiving and signing for "money", even though they noted the lead seal was broken. The guards assumed it had taken a knock in flight, as had happened in the past.

The 68 kilo weight stated on the manifest for the trunk has led Police to the "Skyway Robbery" explanation for the missing money. Also the address provided by the trunk "owners" required by Air Inter in case anything was lost or damaged turned out to be false.

~ ~ ~ ~

This story occurred shortly after Delta Airlines started operating into Hong Kong.

DELTA 419: "Hong Kong Ground, Delta 419-Heavy, we-all got two burnin' and we're ready to roll. Over."

HONG KONG GROUND (Chinese accent-exasperated, yelling ...): "Delta 419, you have two burning? Confirm!"

DELTA 419: "Yep ... two burnin' and ready to roll. Over."

HK GROUND (several steps up the anxiety ladder now ...): "Err ... Delta 419, do you need assistance?"

DELTA 419: "Ahh ... err ... Delta 419, we-all got two turnin' and wanna taxi. Over."

UNKNOWN VERY ENGLISH VOICE: "He means two engines started I believe!"

HK GROUND (extremely relieved): "Oh! Delta 419, hold position, approximate four minute delay. Your terminology is all Chinese to me!"

~ ~ ~ ~

A domestic airliner, with the First Officer flying. He performs a very disappointing landing, and during the taxi back to the terminal, the Captain picks up the handmike, switches to PA and says this:

"Ladies and gentlemen, the Captain speaking, that atrocious landing was courtesy of your First Officer. Fortunately for us all, I will be flying the next sector!"

The First Officer was to say the least, horrified. However, a couple of months down the track, the same Captain and F/O are flying together again. This time, the Captain does the landing, and believe it or not, it is equally bad or worse than the First Officer's landing months before. And yes, you guessed it ... the First Officer picked up the mike, and makes the following PA announcement:

"Ladies and gentlemen, the First Officer speaking. That woeful arrival was courtesy of the Captain. Fortunately, I will be doing the next sector!"

The Captain can't believe it and says ... "What the hell did you do that for?"

F/O: "Well, remember a couple of months ago, after a bad landing of mine you did the same thing?"

CAPTAIN: "Yes ... but I didn't key the mike!!"

~ ~ ~ ~

This is the story of a controller at LAX (Los Angeles) who works in the Tower, and a US 747 flying to Honolulu, which accidentally taxied away from its gate ... without its load of passengers!

It seems the pilot called Los Angeles Ground requesting permission to push back and was given ... "pushback approved, taxi to Runway 24 left and hold."

After the tug pulled the aircraft out from the gate and disengaged, the pilot taxied a few metres, then came to an abrupt halt, and immediately requested clearance back to the gate.

"Do you have a problem Sir?" the controller enquired anxiously.

"Yep", the Captain replied coolly, "we just forgot the passengers!"

~ ~ ~ ~

A small dog wandered onto the active runway at Sea-Tac (Seattle-Tacoma) International Airport and proceeded to hold up departing traffic for about 15 minutes while ground crews tried to catch the little canine.

The passengers aboard the first aircraft in line watched with glee as the dog led all his highly trained would-be captors on a merry and wild chase, until they finally managed to ensnare him. When the Tower controller finally reported that they were cleared for takeoff, the Captain said over the leading aircraft's PA to one and all ... "Dog gone, we go!"

~ ~ ~ ~

At a little outback aerodrome in Queensland a Chieftain was being readied for takeoff.

Before the main flow of passengers boarded, an elderly blind man was wheelchaired and helped onto the aircraft followed by his faithful labrador, which sat right at the front of the aircraft. The pilot was going through his preflight checks and noticed the dog becoming agitated.

After continually whimpering and shifting, the pilot realised that nature was calling and the dog wanted to find somewhere to relieve itself. The pilot donned his pilot's jacket with wings, put on his dark Ray Ban aviator sunglasses to combat the midday sun and carefully led the dog off the plane. Just at this moment the passengers started to walk out to board the plane. The dropped jaws and dumfounded looks by the passengers was a sight to behold.

Piper Chieftain

~ ~ ~ ~

This took place at the Amberley RAAF Base with F-111s and the odd Mirage in attendance, along with some US aircraft which were visiting.

All was fairly quiet until an Ansett 727 arrived in the circuit area. A rather posh voice announced that the aircraft crew would be performing some touch-and-go landings for pilot training.

The 727 came in for what appeared to be a normal approach, but hit the runway with a very solid thump and a good deal of smoke from the tyres. Suddenly the engines were placed in reverse thrust and the brakes applied heavily. The aircraft came to a fullstop in the middle of the runway, and for what seemed like an eternity no doubt to the poor ruffled soul in the left hand seat, there was complete silence on the radio. Then the posh voice reappeared again and said simply ... "We shall now try that again but this time with a real pilot!"

The aircraft then backtracked to the threshhold and took off, did one textbook touch-and-go, and then disappeared into the wild blue yonder!

~ ~ ~ ~

Overheard at San Antonio International Airport a while back, on an IFR day, with drizzling rain. There was three quarter mile visibility and a 500ft ceiling. Only two aircraft are in motion; a Cessna Skylane taxying out and a Braniff 727, in the midst of a pushback. Clearance delivery and ground control are on the same frequency. The Skylane pilot had just read back his clearance to Kerville, Texas, which included "climb and maintain 5000ft and maintain runway heading until advised by Departure Control". A young sounding voice came on saying ... "he probably doesn't know how to maintain either a heading or an altitude, if he even knows the difference!"

The Skylane pilot punched his mike button and said ... "Ground, you can tell that overpaid kid aerial bus driver that this old A&E mechanic was in fact maintaining headings and altitudes before he could maintain clean underwear! And furthermore, I didn't have to hire out to log up flying time. I paid for it, out of my own pocket!"

There was a L-O-N-G silence. Then an older voice.

"You tell 'em dad!"

(After which the ground controller announced that as far as he was concerned the entire subject appeared to be closed.)

Cessna 182 Skylane

~ ~ ~ ~

Monitored late one Thursday evening on Sydney Oceanic Control. The aircraft was a Hawaiian Airlines DC-8.

HAWAIIAN 1452: "Sydney Control, Hawaiian 1452, we're out of 9000 climbing to flight level 370."

CONTROL: "Hawaiian 1452."

(Break of 10 seconds, then ...)

HAWAIIAN 1452: "And err ... Hawaiian 1452, your tops are at 6000, the cloud cover is very thin, from 5000 to 6000. And above that Sir, there's a full moon, and it's a bee-oot-ee-ful night!"

CONTROL: "Right ... well have a good trip."

HAWAIIAN 1452: "Okay, and as we say in Hawaii, A-L-O-H-A!"

~ ~ ~ ~

This story was related at a recent Mangalore airshow, to do with the Kingaroo Glider School.

A student and instructor were in a preflight briefing when someone spotted a carpet snake enter their glider. Being afraid of snakes, the instructor searched the aircraft and eventually found the intruder, about halfway up the wing, on the inside. The wing was subsequently removed and the snake evicted. After a 10 minute rest, the instructor and first time student took to the skies. After being set adrift from their tow plane, at their set altitude, the instructor felt something brush against his neck from behind!

Taking absolutely no chances, he told the student to take the flight controls and hold the glider in straight and level flight. He explained that he was intending to open and close the hatch very quickly and "there may be a sudden pitch change of some significance".

The next time he felt the object brush against his neck, quick as a flash, he opened the hatch, reached behind his head and after grabbing the object behind him, without a look, he threw it out of the glider cockpit. Down ... down ... down!

Once on the ground, one of the club officials notified that he was looking for his cat. When the instructor went back to the glider, he noticed a large amount of cat fur on the dash behind his head. The cat was never seen again, presumed lost in glider airspace!

~ ~ ~ ~

It was early summer, daylight saving had just restarted and the time in Canberra was about 12.30pm. In the Sunshine State it would have been only 11.30am. The players: VH-TJU, at that time a TAA DC-9 jet inbound to Canberra from the north, and Canberra Approach.

The conversation:

VH-TJU: "Canberra Approach, good morning (very chirpy), this is Tango Juliet Uniform, on descent to niner thousand, visual, received Delta."

APPROACH: "Tango Juliet Uniform, Canberra Approach, good afternoon (stresses) descend to seven thousand."

VH-TJU: "Tango Juliet Uniform, roger, seven thousand, and er, (slightly miffed) you're right, it's *good afternoon!* You must remember though, we're a Queensland crew!"

APPROACH: "Tango Juliet Uniform, in that case, *good yesterday!*"

~ ~ ~ ~

Getting back 'Stateside' an incident occurred the day before Thanksgiving, which is the busiest flying day of the year in the USA. A Delta Airlines L-1011 TriStar from Chicago O'Hare is bound for Dallas.

An American Airlines DC-10 was due to depart exactly one minute later on the same route as the L-1011. Freezing rain earlier in the day had jammed O'Hare (ORD) and all flights were at least 30 minutes behind schedule. Ground Control was certainly not bored and understandably not in particularly good humour. The following conversation was heard right through Delta's passenger headphone system:

DELTA 416 (in southern drawl): "Ground, Delta 416, holding short, Bravo 6 intersection, taxiway Echo."

AMERICAN 672 (clipped Brooklyn accent): "Ground, American 672, also holding short, Bravo 6, taxiway Echo Echo."

ORD GROUND (hurried voice): "Delta 416 and American 672, which of you is scheduled to depart earlier and who is in front of who at Bravo 6?"

BROOKLYN ACCENT: "Ground, Delta 416, American is in front of me, he can go on ahead!"

ORD GROUND (in upset tone): "Delta 416 Ground, cleared to taxi runway 32 right, hold short, tower 118.1. Break break, American 672 Ground, cleared to 'penalty box', for holding via taxiway Echo. Expect engine start time in 90 minutes, or when you grow up, whichever later! By that time, you will have explained to your 300 souls on board why they'll spend Thanksgiving at Dallas-Fort Worth, and not with their families, over."

That was probably the last time that American Airlines pilot ever tried to do an impersonation!

~ ~ ~ ~

Frequency 121.7, Sydney Ground, on a Saturday morning. The mood in the Tower was jovial as Continental 008 moved off the International Apron.

CONTINENTAL 8 (female): "Sydney Ground, good morning, Continental eight heavy, taxi?"

ATC: "Continental 8, clear to taxi, hold short runway three four. Good morning Dorothy!"

CONTINENTAL 8: "Continental 8 roger, and it's Jeannie."

ATC: (presses press-to-talk, but is speechless)

CONTINENTAL 8 (male voice): "We get more progressive every day."

ATC: "So I see!"

Later ... Qantas 28, taxying out, with an American male pilot working the radio communications.

AMERICAN MALE (obviously QF28): "Jeannie, you still with us?"

AMERICAN FEMALE: "Affirm."

AMERICAN MALE: "You from Hawaii too?"

AMERICAN FEMALE: "That's right."

AMERICAN MALE: "How come Continental gets all its best pilots from Hawaii?"

AMERICAN FEMALE: "How's that?"

AMERICAN MALE: "Well, Dorothy's from Hawaii too."

(Not very tactful, that last remark! No more was heard from Jeannie or her friends. Dorothy, it seems, had made a very BIG impression on this side of the Pacific!)

~ ~ ~ ~

A Concorde, on departure from Perth, remembering the loss of a section of rudder on the Auckland/Sydney sector of its round the world trip a few years back.

APPROACH: "Speedbird Concorde, contact Perth Control 122.4."

CONCORDE: "Cheers", then ... "Perth Control, Speedbird Concorde passing through 4000 and we would appreciate continuous climb to flight level six-zero-zero if possible."

CONTROL: "Speedbird Concorde, goodday Sir, cleared for continuous climb to flight level six-zero-zero."

ANSETT WA F28: "We bet that made your day Control!" (referring to the flight level which meant sixty thousand feet).

CONTROL: (sarcastically) "Yes, we're all very excited here!"

UNKNOWN: "Yes, but is he going to fly quandrantral?"

ANOTHER: "Doesn't really matter, he hasn't even got a rudder!"

Aerospatiale/BAe Concorde

British Airways

~ ~ ~ ~

This happened a few years ago, while the Ok Tedi gold and copper mine was being developed in PNG. All the aircraft operated by the Missionary Fellowship have 'Mike Foxtrot' as the first letters of their registration. This conversation was overheard by a large number of chopper pilots supporting the development of the mine.

MIKE FOXTROT ALPHA: "Lae Tower, Mike Foxtrot Alpha."

LAE TWR: "Mike Foxtrot Alpha, Lae Tower, go ahead."

MIKE FOXTROT ALPHA: "Ah, Lae Tower, err ... what's your weather like? We're in heavy rain, and I'm not sure where we are."

LAE TWR: "Mike Foxtrot Alpha, Lae Tower, broken stratus, and down towards Salamea, patches of low cloud, and rain advancing."

MIKE FOXTROT ALPHA: "What about Finchaven?"

LAE TWR: "Heavy showers, wind southeast, 20 knots, go ahead your position and ETA Lae."

MIKE FOXTROT ALPHA: "Ahh ... Lae, we're circling a village, no road near it, and (voice going up an octave) hang on, I can see the coast now" (voice fades away).

LAE TWR: "Mike Foxtrot Alpha, Lae Tower, I say again, go ahead your position and ETA Lae."

MIKE FOXTROT ALPHA: "Lae, Mike Foxtrot Alpha, we're now over the coast, and there are a couple of yachts sitting in the bay, and some buildings with tin roofs."

UNKNOWN FIRST VOICE: "My God, what kind of a position is that?"

UNKNOWN SECOND VOICE: "Missionary."

~ ~ ~ ~

DATELINE: 31,000 feet, on board a Qantas flight, traversing Australia, enroute to London.

QANTAS 1: (believing he's on intercom, but he's actually hit the wrong button and is on the air on 128.2) "Good afternoon ladies and gentlemen, welcome aboard your Qantas service to London Heathrow, I'm your First Officer and I'd like to alert you that in about two minutes, you should be able to look out of the windows and see the rabbit fence, the border between New South Wales and South Australia!"

ANSETT 727: "Excuse me Sir, what time are the movies again?"

QANTAS 1: "Oh my God, what have I done!"

CONTROL: "Qantas 1, Sydney Control, don't worry about it. But at the big rabbit fence, call Adelaide Control on 128 decimal 1!"

~ ~ ~ ~

One evening on Perth Approach frequency. VH-BOF, a Royal Aero Club Cessna 172 had departed Jandakot and was heading towards Perth. The student began very confidently and correctly.

VH-BOF: "Perth Approach, Bravo Oscar Foxtrot, passing through 700 on climb to 3000."

APPROACH: (obviously very busy, and in a no nonsense manner) "Bravo Oscar Foxtrot, Squawk ident."

VH-BOF: (again, very confident) "Bravo Oscar Foxtrot!"

But he'd forgotten to release the press-to-talk switch, hadn't he!

VH-BOF: "How the hell do I do that?"

~ ~ ~ ~

A Southern Pacific Nord Mohawk was on final for landing on RW07. Shortly after, this was monitored:

SYDNEY TWR: "Tango Juliet Lima, Nerd on final, line up behind that aircraft!"

Southern Pacific Mohawk

Bill Lines

~ ~ ~ ~

One night after the Brisbane Broncos returned from cleaning up St George in the then NSW Rugby League Grand Final the following was heard:

"Brisbane Broncos Terminal Information Zulu. Runway zero one, Wind zero-three-zero degrees, five to eight knots. QNH 1016, temperature one-niner, cloud one okta three thousand feet. On first contact with Brisbane Ground or Approach, notify receipt of Broncos Information Zulu!"

The Bronco's actual return flight after claiming the Winfield Cup was Ansett 767-200 VH-RMF, which contacted Approach around 8.10pm local.

VH-RMF: "Brisbane Approach, Romeo Mike Foxtrot."

APPROACH: "Romeo Mike Foxtrot, welcome to God's own country!"

Then ... shortly after ...

APPROACH: "Romeo Mike Foxtrot, wind is zero-three-zero degrees, five to eight knots and you're cleared to land. Just for your information, the terminal's crowded with Broncos supporters and the carpark is absolutely chokka-block, so you're going to get one hell of a reception!"

VH-RMF: (Captain) "..eah ... thanks mate, let's just hope the First Officer does a really good landing after all that."

APPROACH: "We can only hope!"

~ ~ ~ ~

An Ansett New Zealand BAe 146 crew at Palmerston North on the North Island of NZ recently found themselves in a 'tight' situation recently ... and passengers were asked to give them a hand to push the jet out of its parking bay, when the crew found they were hemmed in by other close-parked aircraft!

The Ansett NZ 'Whisper Jet' should have landed at Wellington if everything had gone to plan, but it was diverted with about 20 other aircraft into Palmerston North due to fog. The upshot was there were no tugs appropriate to tow the aircraft out and the Captain was reluctant to power up the engines in case the jetblast struck surrounding aircraft. Ansett New Zealand public relations manager John Cordery confirmed that the Captain had requested all able passengers to assist them in their plight to safely push the aircraft out from the bay.

All hands were on deck to push the four engined aircraft clear of the obstructing aeroplanes, in what we're told was a very tight spot. At one stage in the human assisted taxi to the tarmac, there was less than a metre between the 146 and another aeroplane.

Apparently the unusual 'movement' was handled in good humour. I'm just wondering if the passengers all got letters later on requesting them to join the ground crew union!

BAe 146

Peter Clark

~ ~ ~ ~

A certain Qantas Captain recalled the time he operated the Qantas 10 service from London to Bombay, arriving around midnight in Bombay in the middle of a monsoon.

There were thunderstorms surrounding the airport and very heavy rain at the field. The Traffic Controller was starting to get a little bit flustered with quite a heavy workload, due to traffic and weather.

The Tower cleared our correspondent Captain to land, advising that the runway was wet, and the wind was 180 degrees at 10 knots. Gulf Air was on approach, and after Qantas 10 landed, our Captain and his crew monitored this exchange:

BOMBAY TWR: (in sing song accent): "Gulf Air, you're clear to land, the wind is wet at 10 knots."

GULF AIR: "Roger, clear to land, and could we get the wind check again?"

BOMBAY TWR: "The wind is wet at 10 knots."

GULF AIR: "Wind check please."

BOMBAY TWR: "The wind is wet at 10 knots!"

GULF AIR: (now very pukka English accent) "Yes, we've established it's wet, but would you mind telling us what the hell direction it is coming from?"

~ ~ ~ ~

This tale from the Adelaide airwaves involves an F/A-18 Hornet, approximately 40nm (75km) northeast of Adelaide, on Control frequency.

The RAAF F/A-18 was conducting aerial work at high levels, when he was advised of other traffic in his near vicinity, VH-TJJ, a Boeing 737 transiting Sydney/Perth. The conversation went like this, air-to-air.

TESTER 916: "Tango Juliet Juliet, this is F/A-18 Tester 916, how about a little show for your passengers?"

VH-TJJ: "Affirmative sir, affirmative."

TESTER 916: "Tell the passengers to look to starboard, look to the right of your aircraft sir ... and they will see all!"

The F/A-18 pilot let the afterburners rip in the evening darkness, apparently creating a very exciting and unusual display for the intrigued Australian Airlines passengers.

The 737 flightdeck crew were apparently very impressed, requesting soon after that the F/A-18 crew "do it again please", this time for the passengers on the aircraft port side. The RAAF fighter pilot obliged cheerily and shortly after VH-TJJ tracked on towards Perth.

~ ~ ~ ~

A retired Cathay Pacific flight engineer told a story of flying into Russia on a Lockheed L-1011 TriStar. His aircraft was number two to land behind an Ilyushin transport.

On downwind leg they were advised by air traffic that they were "second to land behind the Ilyushin". On base leg they were again advised "second to land behind the Ilyushin". On final approach without warning the Ilyushin crashed. Without any change in his voice the air traffic controller advised that the TriStar was "now number one to land". Perhaps they were used to losing the occasional aircraft in Russia?

~ ~ ~ ~

Cessna 337 Skymaster

Ever flown a Cessna 337 Skymaster, that neat little push pull twin with one engine in the front and one in the rear?

Those who have swear by the breed as it represents a thinking mans answer to the problem of engine out asymmetrics compared to losing one on a conventional twin, which is not always a nice event regardless of what the brochure says about continued flight on one donk.

Anyway this guy is a small time charter operator and owns a 337 up in the north and one day thought he would get the better of four rather smart alec city bankers who were not exactly performing a mission of benefit to the local community. In fact they were in train to evict several farmers from their properties who had been hit hard by drought, falling commodity prices and high interest rates.

Apparently my friend had carefully noted that none of the four had paid much attention to the plane when boarding and most probably didn't even realise it had an engine in the tail. It seems that after reaching cruising height and over predictable rugged terrain, he gently retards the rear engine back to about half power whilst suddenly cutting the front engine back to idle. Add some gyrations, a look of panic and a steep turn or two and you've really got the attention of our city boys who think they are rapidly approaching an appointment with St Peter!

This goes on for about three minutes with the Cessna gyrating about the sky and diving up and down ever closer to the spectacular canyons below. The human cargo is terrified whilst their pilot "fights" to regain control.

Feeling that enough is enough he then pretends to fiddle with the engine controls and lo and behold restores life to the forward engine. Just in the nick of time no less.

Legend has it that the four emerged from the Skymaster white and somewhat less emotive for their crucial meeting with the rural community they had come to bludgeon. Apparently those city bankers never did figure out that it was all a stunt while the pilot nowadays is good for a free beer in every pub in the district.

~ ~ ~ ~

A leading senior golf professional in Sydney who can still break his age (80) off the 'sticks' was a pilot with 463 Squadron RAAF during WW2. Recently he competed in a BMW sponsored golf day which he won. On receiving his trophy he was asked by the BMW representative if he had ever been to Germany. Without hesitation he replied "only at night". It brought down the house.

~ ~ ~ ~

In the early days of postwar civil aviation in Australia things were fairly relaxed and it was not unusual for crew pranks to be developed to a near art form.

Our story has an unnamed airline's DC-3 ambling along between Sydney and Brisbane in the late 1940s. On board is a neophyte flight attendant logging her third day on the run. Not unnaturally, the young woman is rather nervous and a ripe target for the dubious attentions of the two veteran pilots guiding the great white bird this day.

Their plan is to engage the autopilot and open the Douglas' side cockpit windows and hide in the baggage compartment after summoning the FA on the intercom to come up 'in a few minutes". A loose curtain separates the passenger cabin, which is full this day, from the area immediately aft of the cockpit so that the prank can be nicely executed without alarming the all important passengers.

Up walks our eager young flight attendant, who upon opening the cockpit door sees only two moving control columns, open windows, heaps of airstream and engine noise and no pilots!. A scream that could awaken the dead is emitted as she turns and bolts through the curtain to the rear of the aircraft causing pandemonium within the cabin as the startled passengers then witness two white faced pilots clambering from the confines of the baggage area and bolt back into the cockpit.

To our knowledge this was the last time that this particular prank was played out on an unsuspecting FA!

~ ~ ~ ~

Remember that great movie of the late sixties *The Battle of Britain*?

Well, even though it seemed that it was well publicised from start to finish there was always going to be somebody somewhere who missed the 'briefing' and doesn't know that such an epic movie is even being made to celebrate the memory of what was probably history's greatest air war.

The scene opens appropriately five kilometres above the English countryside, the glorious spring sunshine only being interrupted by long lines of fluffy cumulus. From their operating base at Duxford come more than a dozen Luftwaffe Heinkel 111s in typical 1940s attack formation, droning on towards a make believe London, their rendezvous with a squadron of photoplanes and of course their escorting Messerschmitts and the less than friendly Spitfires and Hurricanes.

If you were indeed lucky enough to see all of this you certainly wouldn't think it was 1968 and if you somehow stooged your way into the middle of this piston engined armada, emerging from a cloudbank whilst checking your instrument flight talents, in a RAF Canberra, then what would be your reaction?

This actually happened and the poor pilot, who wasn't up to speed with the fact that a movie was being made, quickly burst out over the radio with an uncharacteristic expletive, firewalled the jet bomber and got out of there real fast explaining later to his CO what he and his crewman had seen.

Legend has it that even now, so many years later, that both of them are constantly reminded of that day high above the fields of England and that somehow they hadn't stumbled throughout the Twilight Zone.

~ ~ ~ ~

Our next item comes from Larry Davis' book on gunships in Vietnam where he recalls a mission involving an AC-47 gunship and a C-47 Dakota psyche warfare aircraft.

Monty Python fans will love this one.

"One of the more interesting night missions was when Spooky (the gunship) was assigned to work with a C-47 psyche war aircraft – Gabby to her friends and unofficially a Bullshit Bomber. It was a standard C-47 with a massive speaker mounted in the cargo door and an ARVN trooper constantly rejoicing over the mike about the benefits of the South Vietnamese government. Gabby would orbit at about 2500ft in a pylon turn and begin talking to the 'little guys on the ground', always imploring them not to fire upon the speaker aircraft or great trouble would befall them."

Unbeknownst to the black pyjama crowd, Spooky was also orbiting above them, at about 3000ft in complete darkness and about a quarter turn behind Gabby. Sure enough, the black pyjama boys couldn't restrain themselves and began taking pot shots at the brightly lit Gabby at which point the unseen wrath of the gunship would fall upon them like lava from heaven. The massive onslaught complete, Gabby would retort "See, I told you so".

AC-47 "Spooky"

~ ~ ~ ~

Q: What's the difference between God and pilots?

A: God doesn't think he is a pilot.

And now some of the greatest lies in aviation:

1. I'm from the CAA and I'm here to help you. 2. We'll be on time ... Maybe even early. 3. I have no interest in flying for the airlines. 4. All that turbulence spoiled my landing. 5. We shipped that part yesterday. 6. I'm always glad to see the CAA. 7. I've got the field in sight. 8. I've got the traffic in sight. 9. Of course I know where we are. 10. I'm sure the gear was down.

~ ~ ~ ~

Not so humorous at the time but certainly capable of raising a chuckle when told (much) after the event was this incident relayed some years back.

It is 1942 and a new group of young US Army pilots have just deployed to the remote reaches of the Northern Territory to assist the RAAF in the defence of Darwin. Their rudimentary base is located deep within the scrub and seemingly a haven for everything alive except humans. On top of that, the often exaggerated lethality of the local insect and reptile population has exceeded all measure of reasonable believability following informed and somewhat dramatic 'briefings' by the local Aussies.

If the Japanese, the oppressive weather sickness and loneliness aren't bad enough then there is the mind numbing array of predatory wildlife to contend with. It is thus with great trepidation that a young pilot preflights his P-40 with additional care prior to a short hop to test repairs made throughout recent days.

The Curtiss lifts off from its dusty base and climbs to 20,000ft where the cold air is a gift from God for this young lad from what would at this time be snow covered Montana. However, his state of relative bliss is short lived as in the middle of a loop he finds company in the form of a two metre black snake which drops onto his face and torso. The giant reptile is understandably upset at being roused from its underfloor slumber by such aerobatics and obviously, like the young pilot, distinctly wishes it was somewhere else.

With a pulse rate high enough to send his heart through the windshield the young lad from Montana levels the Kittyhawk and wrestles with the giant snake in a bid to get it out of the plane. Realising that the deadly snake is as terrified as he and one bite and he's history, the quick thinking lad inverts the P-40 and slides back the canopy, with gravity completing the mission.

Needless to say when he finally landed, he climbed down from the aircraft and promptly collapsed. Years later our pilot was able to honestly recall this as his most hair raising mission of the war.

Rumour had it that after that episode there was more time spent preflighting the planes for snakes than checking that the machine would actually fly!

Curtiss P-40 Kittyhawk

~ ~ ~ ~

It is said that by coincidence, not design, that a Liberal, a Democrat and a Labor politician found themselves on the same chartered light aircraft out of Canberra and on the way to Adelaide.

As the plane passed over the Mallee, the Liberal pulled a $100 note from his wallet, opened a window and let the note flutter from his hands, declaring that it would find its way to a farmer in need.

The Democrat, not to be outdone, pulled two $50 notes from his fob pocket, opened the window and dispatched them on their way, saying that his generosity would help two farmers.

The Laborite got to his feet, pulled ten $10 notes from his back pocket, thrust them through a window and said that he would help ten farmers in need.

All of this activity attracted the attention of the pilot.

"If the three of you don't shut up and sit down", he bellowed, "I'll throw all of you out of the plane and help every farmer in need!"

~ ~ ~ ~

We all know that when in a group, footballers can be, well, a little off. Our story concerns how an entire team and their unruly supporters one fine day took on a petite female flight attendant – and lost.

It seems the team was flying interstate for a major round and occupied more than half the economy cabin of the 727. Naturally you can imagine that things are getting pretty noisy as by this stage the boys are well and truly revved up. It's Party Time.

One lad, renowned for putting brawn before brain, decides to shock the approaching flight attendant who is in the process of offering the beverage service. On the surface she is a demure lass – a smiling person that hides a streetwise wisdom of many years of dealing with such unrulies.

The footballer unzips his fly and displays his perceived manhood in a feeble attempt to shock our FA. With his mates snickering at full throttle the ever smiling flight attendant quickly sizes up the situation, leans forward with the very hot coffee urn and, well we don't know if it was momentary turbulence, or what, but she neatly misses or hero's cup by a wide margin!

Legend has it that this particular footballer not only didn't play that key match, much to his long lasting embarrassment, but also was sidelined for the next five. Definitely a case of FA 1 Brawn nil.

~ ~ ~ ~

A number of regional airlines operate turbine powered aircraft with a passenger capacity of around 18 seats.

Because of their relatively small size, these aircraft are generally operated with a crew of just two pilots. As there is no flight attendant to supervise passenger loading and unloading, this job is normally done by the First Officer, who also gives the obligatory safety brief before rejoining the Captain on the flightdeck.

Recently, at the conclusion of such a flight, the female first officer was farewelling passengers on the tarmac of a country aerodrome when she was approached by an elderly and irate female passenger. "Thanks very much for the cup of tea you didn't get me," she complained, "and I hope that you had a lovely time sitting up the front with the pilot!"

~ ~ ~ ~

This story involves ZK-TAK, a Beech Queen Air, operated by Kiwi West Aviation, and New Plymouth Tower in New Zealand. Tango Alpha Kilo is lined up departure one RW23.

ZK-TAK: "Tower, Tango Alpha Kilo, would it be alright to wait here for a couple of minutes to let the engines warm up?"

NP TWR: "Tango Alpha Kilo, Yep ... that's alright."

(pause for a few seconds ...)

NP TWR: "Actually ... Tango Alpha Kilo, you may have to get out and sit on them for a while!"

Beech Queen Air

~ ~ ~ ~

Once upon a time, the old Civil Aviation Authority and the Canberra Rowing Club decided to compete in an annual boat race on Lake Burley Griffin.

Both teams trained long and hard to reach their peak performance. On the big day, the Rowing Club won by a kilometre.

The Authority team was rather discouraged by their loss and morale sagged. Senior management decided that a reason for the crushing defeat must be found and so a project team was set up to investigate the problem and take appropriate action.

It was found that, while the Rowing Club had eight people rowing and one person steering, the Authority had one person rowing and eight steering. Senior management accordingly hired consultants to study the Authority's team structure. For half a million dollars the consultants advised that the team needed to be better co-ordinated so that more effort went into rowing.

The new Authority team consisted of four steering managers, three senior steering managers, one executive steering manager and one rower. A performance appraisal system was set up to give the rower more incentive and he was sent to courses run by the consultants so that he would feel empowered and enriched.

The next year the Rowing Club won by two kilometres.

The Authority sacked the rower for poor performance, sold off the paddles and halted development of a new boat. The money saved was used as performance bonuses for senior management.

~ ~ ~ ~

During his service career a RAAF fighter pilot had a problem in his Meteor jet fighter and prudently decided to land at the nearest airfield – one not used to such elite machinery.

He landed and came to a stop at the end of the runway and for safety reasons elected to shut down the engines and get a tractor to tow the aircraft to a hangar. He felt relaxed at having handled an awkward situation with coolness and that nothing untoward had happened to him or the aircraft. As the engines wound down he spoke to the Control Tower about his tractor requirement. As light rain was falling he wisely kept the cockpit hood closed.

However, the next 20 seconds were not so calm and collected. A fire engine, with red flashing lights and siren wailing, pulled up alongside. Before the pilot could stop them, two men in astronaut like anti fire suits and helmets began spraying foam over the aeroplane, while another similarly garbed giant placed a ladder against the fuselage and then swung an axe through the cockpit hood in front and behind the pilot.

After removing the shattered perspex the fire fighter produced a knife and bending over the fiercely gesticulating pilot, cut through the shoulder and parachute harnesses; as this was happening another energetic fellow with a similar knife appeared on the other side and slashed through the thigh straps and before the unfortunate pilot could further remonstrate, four arms lifted him bodily out of the cockpit!

RAAF PR

Gloster Meteor

~ ~ ~ ~

When on exercise with the USAF a few years ago a story emerged that, well, just had to be American!

Two fighter jocks were transiting the good ol' US of A in a tandem seater and after a few hours (which is a long time for these guys to be in the seat) the guy in the back seat started to get bored.

He went cold mike and twiddled with cockpit dials and knobs, played with the screens but was still bored. He looks down at his lifevest and takes it apart looking at the whistle, the emergency codes and the sea dye etc. That still didn't excite him so he looks down between his legs and sees a dayglo toggle. After this much boredom he seems to forget that it's the inflation toggle of his liferaft.

To cut a long story short he pulls the toggle just enough to inflate his dinghy. Now that's exciting. It starts to inflate and make him quite uncomfortable in his straps, but because the straps only cut across his hips, his legs start to get pushed up in a quite unnatural and unflattering position. He panics and reaches for his survival knife. He blindly stabs around the cramped cockpit finally puncturing the raft.

The front seat pilot, semi dazed by the relentless boredom of the long transit, hears this loud explosion from behind then immediately sees a wall of talcum powder dispersing through the cockpit. Fearing a catastrophic but smoky engine failure and being over the barren southwest, the pilot calls "eject, eject" – the rear seater sees his buddy catapult skyward and fears that with his preoccupation within his own cockpit he hasn't heard the full story.

He thinks if the captain bangs out we must be in some serous trouble! He ejects as well leaving the poor old once serviceable aircraft to sustain unbalanced flight for a while before crashing into the sparse Arizona wastelands.

Needless to say both pilots had a nice two year ground posting to plan some inflight entertainment for their next posting on long haul C-130 ops.

~ ~ ~ ~

It's in most peoples' belief that animals don't travel very well in any sort of transportation.

This was certainly the case one time when an incident occurred in the Kimberley region of Western Australia. It was a short flight from Kununurra to Derby in a light twin engined pressurized aircraft. The aircraft was on charter for two passengers and had a fair load of cargo.

Included in the cargo was a cat, which was in a cage. When the pilot was loading the aircraft he stored the cat in the nose of the aircraft. Flying time wasn't long so the pilot didn't think the cat would have too much trouble flying 'up front'. Unfortunately when they landed and began unloading the aircraft they found that the cat wouldn't wake up or move. When they opened the cage to investigate they noticed that the cat was frozen stiff!

The pilot immediately realized that the nose of the aircraft wasn't pressurized and as they had been flying at 18,000 feet they quickly understood why the cat was frozen stiff.

~ ~ ~ ~

A long time ago a certain Herc driver was tasked with dropping about two dozen SAS paras into a space constrained dropzone at night. This is not an unusually difficult task, but one that can go wrong if the winds aren't what they are forecast, which of course is ops normal.

Anyway, our C-130 is operating out of Pearce and the dropzone is not that far from the base while the flightcrew has the option of flight planning back to Pearce for a comfortable overnight and leisurely departure next day or heading eastward to Richmond where the strong winter jetstreams will, despite the distance, have them home in time for breakfast.

The Hercules lines up carefully and out go the SAS into the night sky. Unfortunately the aircraft is considerably left of track and the Army elite land not on open fields as planned but amongst woodland studded with rocky outcrops – great!

With several injured and very aggrieved troopers waiting to have a 'talk' with the Hercules skipper if they returned to Pearce an instant decision is made by the Herc crew to wind up the Allisons and head east at best speed. Like they say, discretion is the better part of valour!

~ ~ ~ ~

Just thought you might find this piece from *Inside Tourism* amusing (or perhaps not):

A Delta Airlines jet was forced to land without guidance in the dark after the lone air traffic controller at Florida's Palm Beach Airport fell asleep. Police believe the controller dozed off while cleaning a pistol. Pilots of aircraft circling above the airport waiting to land alerted aviation authorities.

Police surrounded the locked control tower with squad cars, sounded sirens and aimed spotlights at the windows, but to no avail. When they finally burst in, police found the controller waking up, shoeless and yawning. A pistol in the process of being cleaned, along with a full clip of bullets, was found next to him.

Firearms are illegal in Florida control towers and sleeping on duty is frowned upon (!!), but police found no evidence of drink or drugs so the controller was released and was still directing traffic at Palm Beach Airport while an investigation proceeded. Far out!

~ ~ ~ ~

Scene: Hobart Airport, Tasmania.

A Qantas Boeing 747-400, VH-OJB, *Wunala Dreaming* resplendent in its red ochre livery of Australian Aboriginal murals and symbols, is sitting on the tarmac, about to be invaded by hordes of visitors to the airport. It's open day, and the 747 is one of the key exhibits.

Enter, Stage Left: two RPT aircraft into the circuit pattern and they commence a three way exchange of view with the Tower.

RPT JET No 1: "Oh yuk, what's that down there?"

HOBART TWR: "I think they're having a corroboree"

RPT JET No 2: "I suppose that's what happens when you park your plane in the open overnight!"

Boeing 747 "Wunala Dreaming"

~ ~ ~ ~

Legends are made this way while some others just accumulate a mystique that will live with them for always.

Indonesian airlines are often, perhaps unfairly, lumped in that basket as the following item reflects.

A couple of years ago a tourist was in Bali. At breakfast one morning one of the guests said that she was going over to Lombok that day and that she was flying. Another guest said that she was taking a great risk, to which she replied in song: "It's Merpati, and I'll die if I want to."

~ ~ ~ ~

While English is the international language of aviation, not all who attempt to use it are as eloquent as those of us who speak the language at home.

One of the many things with which an 'international' pilot must contend is the diversity of interpretations and accents applied by others to his mother tongue.

Some of these can be quite humorous, while others can have significant safety implications.

One way in which a pilot can be prepared for what he may hear is to try and think like the controller concerned. This may even involve mentally applying the local accent to location names etc so that they are more easily understood when heard. In trying to think with an accent, there is also the risk that you will at some stage try to talk with it too. More than one pilot, including myself, has fallen into this trap.

This is exactly what happened to the Captain of a RAAF Hercules on its way across India to the Middle East. The aircraft was on descent to Bombay where it would refuel before continuing on to Bahrain. On this particular leg, the copilot was flying the aeroplane, so the Captain was doing the normal copilot duties, including operating the radios. He was also, quite unconsciously, developing a very pronounced Indian accent.

Each radio call received from the Indian controller would be acknowledged with a Peter Sellers like parody of the man's own accent. This went on for some time, with no comment from the controller concerned; that is, not until after the aircraft had landed and was taxying to its parking bay.

As the crew were completing the after landing checks, the tower controller got his own back with the following message:

"Aussie 123 – After you are shutting down, please be having the copilot report to operations – and he had better bloody well be velly black!"

The crew were certain that the controller was only kidding. But for some reason the Captain found it positively necessary to personally supervise the refuelling while the others went to operations for flight planning.

~ ~ ~ ~

A true character of our wide skies was F C (Chris) Braund. In fact his initials once served as the registration on a Mustang in which he spiritedly enlivened the airshow circuit for some years.

Many of his priceless verbal sallies were enhanced, if not dependant upon; an engaging speech impediment that was more a hesitation than a stutter. Here are a few typical FCB tales:

Scene One: FCB is just past Gunnedah enroute to Moree in northwest NSW and attempting a position report, dragging it out more than can be conveyed in print.

"Ahhh Ssssydney, Fffoxtrot Charlie mmmmmmmmm B ... ravo, w ... as G ... unnedah at F ... ive eight, eight th ... ousand f ... ive ahh hundred, ahh estimaate N. ... arrabri at z..ero ah d.d..ammit, w..w..e're there."

Scene Two: Back of Bourke where FCB is attempting to pass an ops normal report despite poor HF conditions and much yabber on the only usable frequency. Sydney Flight Service has been trying to raise its opposite number at Lord Howe Island and vice versa on a matter of apparent urgency with both operator's voices becoming more strained as they attempt to be heard:

SYDNEY FS: "Lord Howe, Lord Howe, this is Sydney, this is Sydney on eight eight."

LORD HOWE FS: "Sydney, Sydney, this is Lord Howe, this is Lord Howe, do you read Lord Howe?"

This fruitless exchange takes place many times. Finally a droll distinctive voice penetrates the airwaves: "L..L..Lord ahh ahh Howe I wish you'd sh..ut up".

On another occasion Chris was flying an East-West Airlines Avro Anson (VFR only) from Tamworth to Sydney only to turn back well before reaching his destination. Next day I asked him why he chickened out.

His reply was to the effect "Well John, the w-w-weather was not the b-b-best, and after p-p-passing Mur-mur-murra – oh that place (he could never get his tongue around the position report at Murrurundi) I saw a t-t-train come out of a cloud, s-s-so decided it was t-t-time to go home".

~ ~ ~ ~

Our story, which we believe is true, occurred during the sixties at Avalon airfield in Victoria during the Mirage production era.

It seems that a new Mirage was taxying into position at the end of Avalon's long runway for a routine predelivery test flight. Unbeknown to the pilot and certainly well out of sight was a farmer on a tractor cutting the grass alongside the bitumen near the far end of the three kilometre runway.

Everything is 'go' in the Mirage. The pilot firewalls the throttle and the Atar goes to full burner for a spectacular takeoff that will see the aircraft retracting its gear promptly and maintaining straight and level flight till the end of the runway before pulling up into a high speed zoom climb. Well you can guess what is going to happen!

Totally oblivious to our farmer on his noisy tractor, the Mirage thunders past our hero like a scalded cat at better than 300kt (555km/h) and not that many metres distant. Not even seeing the aircraft or understanding for one nanosecond his fate, the farmer is propelled from his seat by the not unsubstantial shockwave, waking up moments later in the long grass wondering what on earth happened!

Defence PR

Dassault Mirage

~ ~ ~ ~

An airline crew was overnighting in Perth, and crowded into the lift at the hotel. Another person squeezed into the lift as the doors closed, and accidentally hit one of the lady flight attendant's breasts with his elbow. He said:

"If your heart's as soft as your bosom, you'll forgive me." To which she quickly replied:

"If the rest of you is as hard as your elbow, I'm in Room 415!"

~ ~ ~ ~

Passenger to attendant:

"A Vodka and Tonic, please."

Attendant: "Certainly Sir" (goes and brings back the drink).

Passenger: "I'm terribly sorry, I meant to ask for a Scotch and Dry."

Attendant: "No problem, Sir" (gets the new drink).

Passenger: "I'm really sorry, it must be jetlag, but I really meant to ask for the Vodka and Tonic."

Attendant: "No problem Sir, but would you mind first nibbling on my ear?"

Passenger: "Not at all, but can I ask why?"

Attendant: "Well, when someone's screwing me around, I'd like a bit of passion to go with it!!"

~ ~ ~ ~

Tumult in the Clouds is a novel by WW2 fighter ace James Goodson. Goodson, with 32 kills to his name, flew with the original Eagle Squadron in 1940 before transferring to the US Fourth Fighter Group. His book is an excellent account of a fighter pilot's life of the period and contains a plethora of memorable experiences that tell it like it was. One lighter event that is guaranteed to raise a smile though concerns Goodson and First Lt Ralph Hofer who because of his youthful looks was nicknamed "The Kid":

One evening (Goodson writes) I was breaking in a new boy, Ralph Saunders, and flying on his wing, when I saw another P-47 pulling up on the other side of him. It came in close and sitting in the cockpit sat a large Alsatian dog – and no sign of anyone else. The plane then dived ahead of us, pulled up into a perfect loop and ended up behind us.

When we landed, we both climbed into the jeep, and without a word, drove off to 334 Squadron dispersal. The Kid and his dog came over and clambered into the back.

"Duke's getting real good at flying!" said Hofer.

"Any damn fool can do a loop," I said, "but how is he on instruments?"

"Great!" said the Kid.

"Sounds like a real show off," I said.

After we dropped them off, Saunders could not restrain his curiosity any longer. "Sir, did he have that dog on his lap?"

"Yeah," I said sourly.

"That's fantastic." Then sensing my mood, he added quickly, "But I guess you disapprove."

"To get enough room, he leaves his parachute off," I said.

During those long evenings when I drilled my new proteges, I often glimpsed a lone Thunderbolt cavorting among the clouds, diving down through the white valleys between them, pulling up vertically until it stalled and fell off into a spin, pulling out at the last moment and zooming back up in a glorious Immelmann or loop and rolling off the top. And after landing I would drive around the perimeter track to pick up the laughing Kid and his happy dog.

At 21 years of age with 27 victories to his credit The Kid finally met his match. While escorting a bomber raid over Budapest in 1944 he fell victim to one of the greatest fighter aces of all time, Erich "Bubi" Hartmann. The ironic thing was that Bubi is roughly translated as Kid in German and that Hartmann was given his nickname for all the same reasons and was also 21 years of age at the time.

Back at Debden the waiting ground crews knew the instant that the Kid had been shot down for Duke, the big Alsatian, leapt up and let out a series of doleful howls and then went and laid down in the empty revetment where Hofer's Mustang would never be parked again.

Republic P-47 Thunderbolt

~ ~ ~ ~

In the immediate postwar years the Australian air routes were dominated by Douglas DC-3 and DC-4 aircraft.

These, and the Convairs that joined them in 1948, had no public address systems and passengers were informed as to who was up front and who was serving the coffee by nameplates fitted in a holder on the cockpit door. Crew rostering in those days was fairly free and easy in comparison with the complex seniority based bidding system devised for a later generation of fliers. Consequently, an unofficial diversion in those more innocent times was to concoct interesting combinations of pilot and flight attendant names.

Not unexpectedly we could create a reasonable spectrum of colours, including Black, White, Grey, Brown and Green and a Captain Pepperday made a good pairing for a Miss Peppercorn. At one time we had FAs Milde and Wilde and a Captain Savage suitable for casting with a Miss Gripp.

Our finest effort was probably for the inaugural visit by a Convair to the Tasmanian ports of Wynyard and Devonport, when we were able to arrange for FAs Audrey Wynyard and Pat Davenport to serve in the cabin.

It is hoped that our passengers enjoyed our feeble attempts at humour.

~ ~ ~ ~

We'll call this next pair Joe and Pete.

Both gentlemen have Private Pilot's Licences, and on this Sunday morning, Joe was in command, occupying the left seat. As they approached the field, Joe allowed Pete to fly the descent and join up with the circuit.

Joe was listening to another aircraft in the circuit, a Cessna 152, and he recognised the voice of the CFI (Chief Flying Instructor) on board. Joe made the call joining crosswind, with Pete still flying, and watched the Cessna 152 with the CFI slot in behind. Joe said to Pete: "Make it a neat and tidy one mate, the CFI's behind us!"

Then to Joe's horror, he realised the transmit light was still illuminated, and his press-to-talk (PTT) switch had been jammed on as he warned Pete about the CFI being behind them. He thought carefully, and realised they hadn't used any colourful language. As soon as the button was freed, the CFI on the Cessna 152 replied: "I'll watch your landing with great interest!"

The moral to this story. Always check your transmit switches before you talk idly in the aircraft. You never know who could be listening.

~ ~ ~ ~

Most US aircraft callsigns are flight numbers or numerals and some US controllers are not too well versed with the phonetic alphabet. With the foregoing in mind, the following conversation took place between an Australian registered Falcon bizjet approaching a very busy Denver Stapleton International Airport.

(Without pausing to draw breath ...)

ATC: "Continental 752, descend and maintain four thousand, 170 knots please, Continental 728, you Sir, descend and maintain five thousand, United 127, I'll have you contact Approach two-four decimal seven-five, Victor ... err ... ahh, err, Goddam! THAT AIRCRAFT WITH ALL THE LETTERS IN YOUR CALLSIGN, descend and maintain three thousand, contact Approach two-four decimal seven-five, Continental 743, 180 knots please ... " (first breath taken here).

~ ~ ~ ~

Fire authorities in California found a corpse in a burnt out section of forest while assessing the damage done by a forest fire.

The deceased male was dressed in a full wetsuit, complete with dive tank, flippers and face mask. A post mortem examination revealed that the person died not from burns but from massive internal injuries.

Dental records provided a positive identification. Investigators then set about determining how a fully clad diver ended up in the middle of a forest fire.

It was revealed that, on the day of the fire, the person went for a diving trip off the coast – some 20 kilometres away from the forest.

The firefighters, seeking to control the fire as quickly as possible, called in a fleet of helicopters with very large buckets. The buckets were dropped into the ocean for rapid filling, then flown to the forest fire and emptied.

You guessed it! One minute our diver was making like Flipper in the Pacific; the next he was doing breaststroke in a fire bucket 300 meters in the air. Apparently, he helped extinguish exactly 1.78m (5 feet 10 inches) of the fire. Some days it just doesn't pay to get out of bed.

~ ~ ~ ~

The following item came from a recent edition of *Aviation Week and Space Technology*.

Some fellows on a supply flight over the Pacific had become conscious of an annoying buzzing sound coming from a couple of depth bombs in their ammo cargo and, not liking it a bit, they contacted the base for advice. The substance of the reply they got was, "think nothing of it". But they kept thinking of it anyhow, and finally they muscled the two 500 pounders out into space – where, even before hitting the water below, the charges burst with a sky rending WHAM-M-M-M! Five minutes later, there came a call back from base. "Change in earlier message regarding depth bomb ... Further consideration indicates defects ... Jettisoning of charges is suggested."

~ ~ ~ ~

One particularly amusing scene took place when Rolls-Royce was carrying out performance work at 36,000ft in its Dart turboprop powered Douglas Dakota.

Being unpressurised, the plane's interior not unnaturally iced up and resembled a failed igloo so that the only view of the outside world was through a five centimetre opening in the side window of the cockpit.

Through this gap the crew observed a Sabre jet fighter with Canadian markings, the pilot looking sideways in amazement as he slowly slid past before vanishing earthward, having stalled his aircraft in the process. An hour after landing the test pilot received a call from the OC flying at RAF Luffenham: could he please confirm a statement by one of his officers who was offering high odds to anyone willing to bet that he had encountered a 'Gooney Bird', (the affectionate American term for the C-47), at such a ridiculous altitude?

All attempts at a personal side bet were of no avail as a relieved squadron commander learned that his man was not hallucinating.

Dart powerd Douglas DC-3

~ ~ ~ ~

The Shorts 360 is known among Air Traffic Controllers colloquially as a "Shed". Here are some other ATC nicknames: Airbus A300 – "Truck", Airbus A320 – "Scarebus", Boeing 727 – "Thumper", Ansett 737 – "Milk Bottle" (due to the white fuselage and blue tail), Jetstream – "Fruitbat", Cessna Citation – "Slowtation", DHC-6 Twin Otter – "Twotter", Britten-Norman Islander – "Bunny", Piaggio – "Pig", Mitsubishi Mu-2 – "Rice Rocket", Westwind 1124 – "Israeli Bomber", Cessna 172RG Cutlass (retractable gear) – "Gutless", and the Mohawk 298 – "Pterodactyl". You have to keep your sense of humour in gear in the Tower!

~ ~ ~ ~

This beautiful little radio exchange comes to us from Stephen Barlay, who wrote a book titled *The Final Call*. The star is a jet jockey calling his military Tower Controller.

FIGHTER: "Tower, this is chrome-plated stovepipe, Bearcat Triple-Nickel Eight-Ball, Angels eight, five in the slot, boots on and laced. I wanna bounce and blow!"

TRANSLATION: "Fighter jet, callsign 'Bearcat 5558', at 8000ft, five miles out, on approach, gear down and locked, request touch-and-go."

TOWER: "Roger, you got the nod to hit the sod!"

It could only happen in America ... thankfully.

~ ~ ~ ~

Dateline: Brindisi, Casale. A British international airline is the focus of attention. The very pompous, superior-type gentleman we sometimes hear on UK based airline flightdecks comes up on air:

AIRCRAFT: "Brindisi Control, this is Silverbird six-six, good evening to you Sir!"

BRINDISI: "Buona sera Sir, you're cleared to fly level 370."

AIRCRAFT: "Thank you very much Brindisi, err ... and as we're going southbound all the way to sunny Africa, where we shall endeavour to indulge in some local style Christmas jollifications, we'd like to extend our compliments of the season to all your staff at Brindisi Control, and take the opportunity to thank you for your help and co-operation all the year round!"

(A long, pregnant pause follows ... then ...).

BRINDISI: "Say again please!"

The British crew obliged in full, followed by ... .

BRINDISI: "Oh ... what problem Sir?" (then a cascade of frantically-spoken Italian) "YOU DECLARE EMERGENCY ... YES?"

AIRCRAFT: "Oh never mind Brindisi, Flight Level 370, roger ... out."

~ ~ ~ ~

It was a very wet day and rain is more commonly measured in feet in the Gulf of Papua.

Our man has been out in this very marginal VFR soup all day. In and out of small muddy places, carrying curious mixtures of muddy people and strange smelling produce in bound up bundles and bags. At 1500 hours, he taxied up to the big tin shed. This was his third call of the day and the Shorts Skyvan was filthy as was the pilot, who ducked into the shed and after squishing around in shoes full of mud and water decided to discard them, cursing the rain and now squally wind that chopped in from the south east.

At 1540, wearing only shorts and shirt he checked that the load of crocodile skins was securely tied down, taxied out into the takeoff position, did the usual run-ups and checks and sped off into the dreadful conditions. Approaching V1 and with both pilots peering out into the murk they were specially eager to feel the wheels leave the ground.

A sudden scream of pain filled the cabin. The copilot shot a quick glance across to the left seat. The pilot grimaced in obvious agony, uttering a string of obscenities that couldn't be printed here. It's about this time that a natural order of priorities comes swiftly into the scheme of things. Pain, self preservation, to rotate or not to rotate, keep the bird on the hard bit between the cones, more pain. This poor bloke has tonnes of that "I can make life saving instant decisions faster than a flash" ability that we all know good pilots are born with.

The Skyvan is beautifully brought to a swooping swerving stop. By this time as you could appreciate the copilot is more than just a little curious as to what is going on. No more had to be said. The pilot leapt out of his seat and now silent but still grimacing in pain, stood looking down at a very bloody big toe on his left foot. He recalled later that it felt like his foot was being pop riveted to the rudder pedal.

And there was the cause of this 30 seconds of near disaster and blood and pain. The left seater was amazed to look into the angry beady little eyes of a two foot baby crocodile, who having got fed up with all the animated rudder kicking during takeoff, had shown his/her displeasure by attaching itself to the nearest offending item – the pilot's big toe.

Fate, luck of the draw, coincidence or divine intervention? The copilot was wearing boots, the pilot was bare footed. How did the angry little reptile get down into the "front office" anyhow? It's said the pilot bled from his wound all that night, but he did heal well within a week.

But a badly bruised ego – well that's another tale indeed.

~ ~ ~ ~

A friend was in the RAF and spent much of his career working on Shackleton long range maritime patrol aircraft. Legend goes that there was a Shackleton out on patrol one day and after another obviously boring sortie they spotted an American aircraft carrier cruising along. So to relieve the boredom and have a bit of fun they decided to carry out a simulated approach to the carrier as there were no air ops in progress.

Now you can imagine the reaction of the Americans on seeing this Shackleton approaching with gear and flaps down, having tracked it on radar for a time but not having actually made radio contact with it, and with one or two engines shutdown, lining up for what looks like a landing on their carrier!

They naturally think it's in trouble and needs to make an emergency landing. So the Americans frantically start clearing the flightdeck, which includes pushing a few

Avro Shackleton

aircraft over the side. The Shackleton then executes a missed approach and continues on patrol and returns to its base.

Rumour had it that the Shackleton pilots apparently did very little flying after that and one can only imagine what was said on board the carrier!

~ ~ ~ ~

This story is about the time a traveller was on a Fokker Friendship flight – it was probably back as far as the Airlines of NSW days.

Peter had done a day trip to Casino and was on the evening flight back to Sydney. Having missed lunch he was very hungry. Dinner was served with the usual knives and forks in a sealed bag which he undid and used. The chap next to him, John, didn't touch his meal and Peter opened conversation by saying "Not hungry mate?" John replied he was a nervous flyer and was so churned up he couldn't eat. Peter responded that he was a regular flyer and generally liked the airline food because he usually managed to miss meals while travelling, as he had that day. John then asked if he wanted to eat his meal as well and Peter was hungry enough to think "why not".

But all they did was swap over the plates with main and dessert and the roll and so John kept his knife and fork etc so all he had of Peter's were the dirty plates and opened butter sachet etc. He still had all his knives and forks in their sealed packet. John didn't give it a thought at the time but later when the hostie (as they were then) came to pick up the plates, she paused, looked at him, then his tray and him again and said "Sir would you like a hand towel?" – no smile, just a dead straight expression.

Peter just cracked up, but John gave him a filthy look and another to the hostie. Peter stopped laughing and the hostie picked up both trays and left without further comment. John was red in the face, jaws and fists clenched. Further conversation was impossible.

If the poor bloke was a nervous flyer, Peter took the prize for being the most unhelpful neighbour he ever had.

~ ~ ~ ~

United 815 from Sydney to Melbourne, early one morning.

UA 815: "Ok Sir, you want us to call control on 125 decimal 3. You have a good day!"

ATC: "I will if the Swans win!"

UA 815 (slightly confused American): "Err ... oh right!"

(For those not familiar with Australian sports, the Swans are the Sydney based AFL team.)

~ ~ ~ ~

During the time of the Vietnam War, No 41 Squadron RNZAF was operating antique Bristol Freighter transports from its base at Changi in Singapore. Operations for 41 Squadron took the New Zealand aircraft throughout South East Asia on transport support tasks, including remote bases in South Vietnam.

One day, one of the bulbous nosed Freighters landed at a small American airfield in the southern Mekong Delta region of Vietnam. The surprised American controller asked "Say, buddy, what type of airplane is that?"

"It's a Bristol Freighter", replied the proud Kiwi captain.

"Goddam", spat out the Yank, "did you build it yourself?"

~ ~ ~ ~

On arrival at Melbourne's Tullamarine Airport from a recent day business trip to Sydney, a businessman flung a small travel case he had taken with him onto the back seat of his RX-7 sports car and left the car park for home.

Later, as he drove through the city, he noticed his tank was nearly empty, called into a nearby service station and asked the attendant to "fill her up".

As the attendant did so he observed the suitcase. "Been travelling?" he enquired replacing the fuel cap.

"Yes," he replied, "just come back from Sydney."

"How long did it take you?" he asked taking the money and delving into his bag for change. The businessman glanced at his watch and noted the time. "Left there two and half hours ago," he answered.

The attendant stepped back sharply, looked at the car, looked at the traveller, the bag, scratched his head in bewilderment and exclaimed, "Cripes mate, you must have been moving."

~ ~ ~ ~

Qantas flew its last 707 trans Tasman service, Auckland/Sydney, in 1979.

Some weeks prior to the flight it became obvious that demand would exceed seat availability as airliner buffs near and far booked for what would truly be a sentimental journey.

Qantas in Auckland telexed head office: "Due heavy demand by aviation buffs for seats on final 707 flight (QF44 AKL-SYD 25 MAR) suggest substitute 747."

~ ~ ~ ~

October 1965, 'Confrontation' is still in full swing in Borneo, and the Blackburn Beverleys of 34 Squadron RAF, were doing invaluable service with daily air drops of food and fuel to forward dropzones (DZs).

The mighty Bevs also undertook the occasional air-land task. On the 29th, XH116 left Labuan for Jesselton (now Kota Kinabalu) to uplift a platoon of Gurkhas to Tawau and then return to Labuan via Sandakan to drop off a platoon of Royal Malaysian Regiment (RMR) personnel.

All went well until start-up and taxi at Sandakan. For a number of reasons, not least the late arrival of both the Gurkha and RMR passengers at their respective enplanement airfields, they were well behind schedule. It was mid afternoon when they left Sandakan – or at least tried to do so! There may have been just a little bit of 'get-home-itis' around as they rolled along the unfamiliar taxiway – and ran the port wingtip straight into a power pole-cum-lamp post about 500 metres from the runway threshold; coming to an abrupt and embarrassing halt in a cloud of dust and a shower of sparks!

The Beverley had just refuelled, so seeing a sizeable chunk of wing hanging loose near the shorting-out power pole, and not knowing if a fuel tank had been ruptured, the crew rapidly vacated first the aircraft, then the immediate area. "Rapidly" may be something of an overstatement, actually, considering that both the Bev's cockpit and passenger boom were seven metres above the ground, and no passenger steps or inter-deck ladders were in place. Nevertheless the Air Movements team certainly would have given the local orang-utans a good run in the speed and agility stakes that day as they scrambled down the inside of the fuselage and left the aircraft through the nearest door.

After the dust had settled, and no immediate conflagration had occurred, they gingerly made their way back to the poor old Bev. She had lost about a metre off the port wingtip, the port aileron was hanging off the outboard hinge, and, yes, the outboard tank was seeping avgas. Clearly '116' wasn't going anywhere for some time!

Nor, with the fuel leak, was anyone keen to reboard the aircraft, crank up the generators and call Labuan to inform them of their predicament. Instead the crew forlornly trooped back to the huddle of airport buildings. Sandakan was then a small airport with limited facilities. As they arrived at what passed for the terminal they met a small, dapper Anglo Indian gentleman in an immaculate khaki uniform leaving the building.

Aircraft Captain: "Can we use your phone, please? We've a small problem".

Airport Official: "Oh dear me no, Sir. We are having no telephone service in the terminal at this time". (Peter Sellers would have been proud of the accent!)

Captain: "Can we use the radio in the Tower, then, please? We really need to contact Labuan urgently".

Official: "I am most exceedingly sorry, Sir, but the Air Traffic Control Tower is having no radio either. Some bloody fool has knocked down a power pole somewhere, and we are having no electricity or telephone whatsoever to the airport".

Most of the crew left the Captain to explain the situation, slunk back to '116', drew lots for who would fire up the system (unfortunate choice of words) and radio Air Headquarters at Lab.

Eventually they returned home ignominiously in the back of an RMAF Herald much later that night!

Blackburn Beverley

~ ~ ~ ~

One contributor tells us that he is a marine radio buff, and became interested in his hobby when he heard the *Queen Mary* passing Gibraltar. The radio signal station flashed the challenge "What ship? What ship?" The *Queen Mary* came back with "What rock? What rock?"

Anyway ... on to our story, monitored on a maritime VHF frequency.

STATION 1: "Please divert your course 15 degrees north to avoid collision."

STATION 2: "Recommend you divert your course 15 degrees south to avoid a collision!"

STATION 1: "This is the Captain of a US Navy ship. I say again, divert Y-O-U-R course!"

STATION 2: "Negative, I say again, divert your course!"

STATION 1: "THIS IS THE AIRCRAFT CARRIER *ENTERPRISE*. WE ARE A LARGE WARSHIP OF THE US NAVY. DIVERT YOUR COURSE NOW!"

STATION 2: "This is a lighthouse! Your call Captain!"

~ ~ ~ ~

Some years ago a select number of Ansett Airlines pilots were dual endorsed to fly Fokker F27 landplanes and Shorts Sandringham flying boats.

They could one day be operating a landplane at Mascot and the next day a flying boat into the Rose Bay water airport. One of those pilots had often been asked, "have you ever considered that you could make a mistake and land a flying boat at Mascot Aerodrome or a landplane at Rose Bay?" His reply was usually "not yet".

On one occasion though a flying boat was returning to Sydney from Lord Howe Island and when 30 nautical miles (55km) out was instructed by the Sydney tower to continue on a westerly heading of 250 degrees thence a right hand turn and alight on a heading of 270 degrees at Rose Bay water airport.

The Captain and the First Officer of the flying boat had operated a landplane service the previous day at Mascot and the Captain being "a bit of a skylark" decided to continue on the westerly heading for Runway 25 and when almost onto the aerodrome turned right and headed for Rose Bay. Whilst taxying back to the mooring the Captain turned to the First Officer and said "Ha, ha, ha, you thought I was going to land at Mascot, didn't you?"

The Captain taxied to the mooring, shut down the engines, put on his cap, picked up his navigation bag, opened the forehead hatch and promptly stepped out onto the water.

The First Officer turned to the Flight Engineer and said, "I think he may need a life jacket!".

An Ansett Shorts Sandringham at Rose Bay

~ ~ ~ ~

In 1968 the RAN fleet tanker HMAS *Supply* left Sydney on its way north to Manus Island with a load of fuel oil to refuel HMAS *Sydney* during one of her trips to Vietnam.

As it sailed up the North Coast of NSW *Supply* had organised for several RAAF Canberras to fly out from Amberley to test out the AAA gun drills of the 40/60 Bofors gun crews. Standing proudly on the Gun Direction Platform, which was a small area on top of the radar shack immediately aft of the bridge and equipped with an intercom system to the forward twin and aft single Bofors, was the Gunnery Officer. He had been newly appointed to the position and was about to leave a lasting impression on everyone who was standing nearby on the bridge.

As the first pair of Canberras came in low and fast from the starboard side, the gunnery officer shouted his orders to the guns via the intercom – including "Two bogies approaching, bearing red 90" – and immediately all guns trained to port. On overhearing the gunnery orders, the Captain realised the error and called out "Don't you mean green 90, Mister Jones?" – and amid the roar of two low flying Canberras passing overhead came the sheepish reply, "I'm sorry sir, but I was facing aft at the time!"

By the time those on the bridge had composed themselves and wiped the tears of laughter from their eyes, the two Canberras were nothing but receding specks on the horizon – probably with two very confused pilots wondering why the ship's AAA guns had been turned away from them.

~ ~ ~ ~

The following comes from a Royal Flying Corps incident report book during WW1 and makes for some interesting reading. There were six avoidable accidents this month, the lead item announces ...

1. The pilot of a Shorthorn, with over seven hours flying experience, damaged the undercarriage on landing. He had failed to land at as fast a speed as possible, as recommended in the aviation pocket handbook.

2. A BE 2 stalled and crashed during an artillery exercise. The pilot had been struck on the head by the semaphore of his observer who was signalling to the gunners.

3. A pilot in a BE 2 failed to become airborne. By error of judgement he was attempting to fly at midday instead of during the best lift periods, ie just after dawn and just before sunset.

4. A Longhorn pilot lost control and crashed into a bog. An error of skill on the pilot's part in not being able to control a machine with a wide speed band of 10mph (17km/h) between top speed and stall speed.

5. Whilst flying in a Shorthorn, the pilot crashed into the top deck of a horse drawn bus, near Stonehenge.

6. A BE 2 pilot was seen to be attempting a *banked* turn at a *constant* height before he crashed. A grave error by an experienced pilot.

After those salient words of advice the report went on to list a number of *unavoidable accidents*.

1. The top wing of a Camel fell off due to fatigue failure of the flying wires. A successful emergency landing was carried out.

2. Pigeons destroyed a Camel and two Longhorns in midair contacts.

Accidents during the last three months cost three hundred and ten pounds, seven and sixpence, enough money to buy new gaiters and spurs for every pilot and observer in the RFC.

~ ~ ~ ~

This was monitored on Melbourne Control just after VH-EAK had requested a climb from FL 310 to FL 370:

MELBOURNE CENTRE: "Ahh, Echo Alpha Kilo, I wonder Sir would you be good enough to show a friend of mine the cockpit. She's on board your flight today, and her name is Madden, I believe she is in seat 15-Charlie. She's a blonde lady."

VH-EAK: "Have you heard how to confuse a blonde?"

MELBOURNE CENTRE: "No."

VH-EAK: "It's easy. You place her in a round room, then ask her to sit in the corner!"

MELBOURNE CENTRE: “Very good, but I’d advise you not to tell my friend that one. She’s a bit sensitive to blonde jokes.”

VH-EAK: “Don’t worry, we’ll look after her for you. What’s your first name Sir?”

MELBOURNE CENTRE: (Name provided but edited from this story)

VH-EAK: “Yes (names controller’s name), we’ll tell her YOU told US the joke! Unless of course we get that higher level.”

(Now that’s one that we haven’t heard pulled before. Very enterprising!)

~ ~ ~ ~

This story was told to me by one of the skeletons from GAF. (Skeletons were the staff who worked in the spare parts area during the Christmas closedown.) An urgent telex (old fashioned fax to you youngsters) arrived from a Nomad operator in PNG.

They had bent their aircraft and needed parts immediately, (pilot error, not another Nomad problem for you cynics). This type of request is called an AOG or Aircraft On Ground and parts generally were urgently dispatched same day via air freight as it was in this case. The next day a telex arrived wanting to know where the parts were. We did a bit of phoning around to the freight company and airline, to be told the parts had already arrived in PNG.

Not wanting to waste any more time, a new set of parts was despatched duly stamped AOG. The next day still no sign of them. This was becoming annoying not only to us losing all these parts but also to the operator who had an aircraft on the ground not producing any income. It was decided to dispatch a third set and have a LAME accompany them. He could also help assemble the aircraft.

On arrival the LAME checked the cargo, and found all six crates present. He was told there would be about an hour’s delay in unloading, so he headed off for what he thought was a well deserved break. Sitting in the canteen drinking his coffee he noticed a truck loaded with his crates heading out of the terminal. Determined not to lose this lot he rushed out, grabbed a taxi, and set off in pursuit.

Eventually he caught up with the truck, jumped out of the taxi and was about to have a go at the driver only to be confronted by a very confused minister from the Assembly of God church (AOG) wondering what he was to do with all these crates of aircraft parts marked AOG urgent that he’d received over the past few days!

GAF Nomad

~ ~ ~ ~

Do you know about the Darwin Awards? – It's an annual honour given to the person who did the gene pool the biggest service by killing themselves in the most extraordinarily stupid way before having offspring. One year's winner was the fellow who was killed by a Coke machine which toppled over on top of him as he was attempting to tip a free can of drink out of it. A recent Darwin Award nominee was ... .

The Arizona Highway Patrol came upon a pile of smouldering metal embedded into the side of a cliff rising above the road at the apex of a curve.

The wreckage resembled the site of an aircraft crash, but it was a car. The type of car was unidentifiable at the scene. The lab finally figured out what it was and what had happened.

It seems that a guy had somehow gotten hold of a JATO unit (Jet Assisted Take Off – actually a solid fuel rocket) that used to be used to give heavy military transports and fighters an extra "push" for taking off from short airfields. He had driven his Chevy Impala out into the desert and found a long, straight stretch of road. Then he attached the JATO unit to his car jumped in, got up some speed and fired off the JATO!

The facts as best as could be determined are that the operator of the 1967 Impala hit JATO ignition at a distance of approximately 4.8km from the crash site. This was established by the prominent scorch and melted asphalt at that location. The JATO, if operating properly, would have reached maximum thrust within five seconds, causing the relatively lightweight Chevy to reach speeds well in excess of 500km/h and continuing at full power for an additional 20 to 25 seconds.

The driver, soon to be pilot, most likely would have experienced G forces usually reserved for dogfighting F/A-18 pilots, basically causing him to become insignificant for the remainder of the event. However, the automobile remained on the straight highway for about 4km (15 to 20 seconds) before the driver applied and completely melted the brakes, blowing the tires and leaving thick rubber marks on the road surface, then becoming airborne for an additional 2.2km and impacting the cliff face at a height of 38m, leaving a blackened crater one metre deep in the rock!

Most of the driver's remains were not recoverable.

Only goes to show ... speeding never killed anybody – but stopping did.

~ ~ ~ ~

During the latter stages of World War 2, the RAAF operated a Pacific island airstrip which was frequently used as a refuelling stage by transport aircraft travelling to and from the US.

Upon the arrival of any aircraft, the most popular spot on the airstrip was of course, the latrine block, which in accordance with normal service practice at the time, consisted of wooden seats over deep holes in the ground all surrounded by hessian screens.

To relieve their boredom in the weeks following the Japanese surrender, some RAAF ground personnel booby trapped the women's latrine with strategically located speakers connected to a microphone equipped airman hidden begind a nearby palm tree.

Favourite victims were American nurses who, on landing, would make a bee line for the "ladies".

Their expressions of wellbeing and relief however, would quickly change to horror when a voice would boom from down below – "Fair go miss, we're still working down here".

The practice was subsequently stopped to avoid injury after a number of nurses tripped during their hasty retreat from the latrine block!

~ ~ ~ ~

Back in the early straight deck carrier days, there was a spate of incidents where Sea Furys were snapping their tail oleos off on deck landing.

Some pilots saw one of these incidents when standing on the "goofers" platform which was located at the aft end of the island. The aircraft landed OK but the tail oleo snapped off and shot down the flightdeck heading for the port side just near the barrier position amidships.

As they watched the oleo fly down the deck, they saw to their horror that it was heading straight for a young sailor who was standing in the aircraft handlers' space with his head just above flightdeck level. The whole thing was happening in slow motion; just like an old movie where it's a toss up whether the hero will get out of the way of the train in time.

In the event the young lad ducked his head as the oleo shot over the side, missing him by a hairsbreadth. The amusing part was to see him slowly rise up out of the sponson looking around like a cartoon character, you could almost hear the words "What's up Doc?"

~ ~ ~ ~

Some tourists were flying with Qantas into Denpasar at the beginning of a four week trip overland through Bali, Java and Sumatra and onto Singapore.

As they were making their final run in for landing they could see a large gathering of people near the airport boundary fence. After clearing the normal formalities their guide ushered them onto a bus for the trip to their "digs" at Kuta and away they went.

A little down the road they passed this crowd that they had noticed on their flight in, "what's the crowd" the group asked. "Just a DC-9 which crashed on landing a couple of hours ago" he calmly explained, "plane broke in three pieces" he went on "but no one was killed" he further assured them.

Needless to say they approached their next four domestic sectors with great apprehension but they had nothing to worry about and lived to tell the tale.

Remains of the Douglas DC-9

~ ~ ~ ~

During a ferry flight with strong headwinds and total overcast between Woomera and Forrest (WA), a Macchi was crewed by an instructor from Central Flying School and a rather senior officer from Air Force office, who had been a fighter pilot of some repute during his earlier days.

After the allotted amount of time in the cruise, the instructor began to get somewhat anxious due to the fact that the Forrest NDB was still very weak on the ADF. Suspecting an increase in headwind as the problem the instructor decided to remain at height until passing the Forrest NDB from where they could make an NDB approach.

After some time and still no overhead indication on the Forrest NDB the Macchi was now very short of fuel. Deciding that all was lost and that it would be smart to save some fuel for landing the instructor decided to use the Macchi's substantial glide range and risk an inflight relight. He then informed his rather surprised passenger of his intention and shut the engine down.

Not to be concerned with this rather rash move the senior officer in his most calm fighter pilot voice said quietly:

"Squadron Leader, I hope for you this is a good career move."

After gliding some 20,000 feet and 40nm (74km), the instructor relit the engine, commenced a climb using the remaining fuel and passed overhead the Forrest NDB from where he let down in a hurry and landed with the most minimal of fuel remaining. Fortunately the instructor survived to see another day.

~ ~ ~ ~

From *The Sydney Morning Herald's* Column 8 (or 'Granny's Column', as we older folk remember it): Just back from a business trip to China is Bruce Martin, of Turramurra. On an internal flight, on China Eastern Airways, he found a chart in the inflight magazine of the characteristics of various airliners, showing their 'Power Plant', 'Maximum Moving Capacity' and Maximum Cruising Fright Level' – whoa, what was that last one?

He was quite happy with the airline, but better spelling may calm the passengers!

~ ~ ~ ~

This story revolves around one of our domestics shipping a black pet cat to a Queensland city.

On arrival at the destination the cat was found to be dead. Out of concern for the owner's feelings, the cargo staff went to the local RSPCA to find a suitable substitute, taking great care to ensure the replacement closely resembled the deceased.

After anxious examination, they decided they had finally found one and duly delivered the animal to its owner. Imagine their surprise when the elderly lady owner, on seeing the cat said, "That's not my cat!" The well meaning cargo officer insisted that it was, as all the documentation was in order, but the lady was equally insistent it couldn't be.

After some minutes of discussion and with increasing anxiety, the cargo officer finally asked how she could be so sure it wasn't her pet, to which she replied: "My cat died last week and I was flying it home for burial!"

Well so much for good thoughts!

~ ~ ~ ~

August 1967 in South Vietnam was marked by one of the few extended visits to the region by an RAF aircraft other than the British Air Attache's Saigon based DH Devon. The two week deployment had an inglorious start, with 34 Squadron Beverley XL150 requiring an engine change deep in 'Tiger Country' on the very first sortie. By the second week of the deployment, however, things had improved. The replacement Bev, XB264, was in full song – amazingly still with the same four engines with which she had left Seletar! The MAMS Team was working hard as always – and they were even talking to the copilot again after his *famous faux pas* with a Green Beret officer.

On the morning of August 24 we left Saigon with a mixed load of rice, blankets, tinned goods, milk powder – and a complete crated bathroom and toilet set, down to the last tile, tub, tap and dunny seat! These were long awaited creature comforts for a group of French nuns operating a convent and mission school at Kontum in the Central Highlands, we were told. We first landed at Da Nang, where we off-loaded some of the rice and blankets, and enplaned two of the Kontum based nuns. Actually, only one of the two was French, a tall, wiry lady in late middle age with a weather beaten face and a lightning fast flow of speech which completely defeated Year 10 (failed) school boy French! Fortunately, however, she was accompanied to the aircraft by an American Peace Corps worker who was fluent in French and, for an American, almost fluent in English too. The other member of the party was a young Vietnamese novice, who, judging by her glazed eyes and open mouthed amazement, had never been near an aircraft of any sort, let alone something the size of the Bev.

The crew filled the space left in XB264's cavernous freight bay by the consignment off-loaded at Da Nang with more cases of tinned and dried food, and with some difficulty assisted the passengers up the side of the fuselage and into the tail boom. (For those of you not familiar with the Beverley, the freight bay was 12m long x 3m square, and though passengers could be carried there in para seats, the preferred pax space was upstairs in the tail boom, where there were 30 comfortable airline style seats, cabin heating [sometimes] and even two toilets with hot and cold running water.

Access to the boom was normally by external passenger steps – if you could find any with a 7.5m reach – or by a demountable inter-deck ladder. Carriage of the latter cut down cargo capacity considerably, however, and the third alternative for pax enplanement was to simply have them climb up the inside of the fuselage, using the cargo bay stringers as a ladder. The older nun went up into the boom like a kid up an apple tree, but the young novice needed a considerable amount of persuasion and help to make the climb; all the more difficult when one is reluctant to employ the usual 'push on the bum' method normally used to assist passengers up the side of the fuselage! With some difficulty and a little embarrassment all round, we got her safely seated, and the Team climbed up into the boom, too, along with one of the Air Quartermasters ('AQM' = RAF for 'Loadmaster').

Finally airborne out of Da Nang, the aircraft climbed steadily up over the Central Highlands. After about 50 minutes flying, the flightcrew started the descent into (what they thought was) Kontum. The crew had been briefed that, though they knew that they were there on a peaceful relief supply mission, this information might not actually have got through to the gentlemen – and ladies – in black pyjamas on the ground, who tended to treat any aircraft, particularly large, slow moving ones (like, say, a Beverley), as targets. At this stage the VC was not thought to have anything as sophisticated as shoulder launched SAMs or any other weapon heavier than .50 cal or 12.7mm. Consequently, the technique was to arrive overhead the (secure) airfield at a height well out of small arms range, and then make a steep spiral descent and land as quickly as possible – no long, low straight-in approaches here!

Just after the aircraft started its spiral, however, we became aware that the two female passengers were very agitated. They were pointing out the window, first at the ground and then further down the valley into which they were descending, all the time keeping up a rapid flow of mixed French and Vietnamese. Most of it was way too fast for the officer in the back, but he didn't need to have studied at the Sorbone to get: "C'est ne pas Kontum! C'est ne pas KONTUM!"

He wasn't on intercom on this particular trip, but by now the AQM, too, was aware of the passengers' concerns, and he relayed his somewhat ad-hoc translation of the old nun's protestations to the front office in a calm scream. Subtle comments such as "Where the **$%%@* are we **going**?", "The old girl says this **isn't** Kontum!" and "What are you **$%%@* blokes getting us **into – Sir**?" quickly had us levelling off at about 8000ft AGL and having a good look round. There *was* indeed an airstrip below us, but for what was supposed to be the centre of US and South Vietnamese forces in the area, it did seem somewhat small, deserted and overgrown! There was also a **much** bigger airfield visible about 10 miles further down the valley, however, which did also appear to have **lots** of aircraft, vehicles and buildings.

We did a quick '180', and within a few minutes were on the ground at the **real** Kontum.

Once on the ground, the aircraft was met by the local Catholic Relief Agency reps, some USAF personnel and a gaggle of nuns from the local Convent, all of whom had been watching the initial descent 10 miles to the north. (You can't do *anything* discretely in a Bev!). Everyone seemed to be talking at once, in English, French and Vietnamese. In the interests of better communications, the MAMS Team added Scots, Geordie, Welsh and Cockney to the discussion. This helped a lot, and it transpired that while the disused airstrip towards which we had been descending wasn't *actually* in VC hands (well, not today at least), it certainly wasn't the place to land 60 tonnes of sitting target (like, say, a Beverley) unless you had a multiple Huey gunship escort! To be fair, what maps we had onboard were pretty old, local nav aids were minimal, we didn't have the right frequencies for most of the Yank comms, and the runway headings at Kontum and the other strip were almost identical ... but still ...!

As with all our arrivals at aid delivery locations during their Vietnam visit, there was a reception committee with speeches of welcome and thanks, and garlands for the crew. Once we were all lined up in front of the aircraft and several lovely young ladies in the local *al dzai* costume had hung strings of frangipani flowers round our necks, the aircraft Captain called the crew to attention. He stepped forward, and with a glint in his eye pulled the Navigator's brevet off his bush jacket (they were only pop studded on for ease of laundry in the SE Asian climate) – and formally presented it to our former passenger, the flying nun!

Now they don't know how they did things in the French Indo-Chinese armed forces, but there was clearly no doubt in the old girl's mind that the Nav had just been stripped of his rank, was about to have his buttons cut off and have his sword – and he had one handy – broken across the Captain's knee! If her speech had been rapid before, it now became positively ballistic as she interceded for the young man's future and perhaps even his life! Eventually, with the help of several interpreters, they got her calmed down and explained that it was only an English joke, and not a permanent end to the Nav's career. Obviously, as far as she was concerned, understanding that it was just an "English joke" clearly explained *everything*.

The rest of the deployment was uneventful, though the Nav completed it without a brevet on his uniform; he hadn't brought a spare half wing with him. The Bevs were out of RAF service by the end of 1967 and 34 Squadron was disbanded on January 5 1968. I would like to think, however, that somewhere there's an elderly retired French nun (do nuns retire?) who still has an RAF Navigator's brevet, and still smiles when she remembers the crazy English crew who couldn't find the right Kontum.

~ ~ ~ ~

A Turkish Starfighter

Paul Merritt

Some years ago, the Canadian Armed Forces (CAF) decided to pay off its fleet of CF-104 Starfighters. Under the Military Assistance Program, they were to be purchased by Turkey.

The first group of Turkish pilots arrived for conversion training in Canada, armed with rudimentary English language skills. Although this proved to be chafing at times, it was generally sufficient for training and Air Traffic Control purposes. One fine sunny day at the CAF base, there were several Turkish manned CF-104s in the circuit area practicing their skills, when a call was received by the Tower:

"Ground – Turkey Starfighter 221, request taxi please."

"Turkish Starfighter 221, Ground, clear taxi RW21, time 43."

"Ground, Turkey Starfighter 221, roger taxi, thank you, RW21."

As the aircraft began a sedate roll to the threshold, the controller noticed more than usual smoke emanating from the aircraft's tailpipe.

"Turkish Starfighter 221, Ground, you appear to have smoke coming from your tail section. Please check and confirm operations normal."

No reply. As the aircraft continued its journey, more smoke was appearing.

"Turkish Starfighter 221, Ground, you have large amounts of smoke coming from your tail section."

Still no response – and now flames could be seen popping out the tail.

"Turkish Starfighter 221, 221, ALERT – YOUR AIRCRAFT IS ON FIRE."

The pilot continued to taxi, unaware of his impending doom, and much to the concern of the now panicking controller who was activating the crash alarm. As the Starfighter continued to taxi past the control tower, smoke and flame continued to increase! The controller now switched to GUARD, the frequency that would reach all aircraft.

"Turkish Starfighter passing Control Tower, you are on fire – repeat – on fire. GET OUT! GET OUT!"

To the relief of the controller, Starfighter 221 promptly braked, the canopy was jettisoned, followed by the pilot swiftly climbing out of the now blazing aircraft. But his relief was short lived as one of the Starfighters in the circuit popped its own canopy, and the pilot ejected – leaving a perfectly serviceable aircraft to plunge into the nearby countryside. The pilot was on downwind adjacent the Tower, had heard the GUARD call and ejected without checking!

~ ~ ~ ~

This story emanates from the protracted war that raged on, largely unnoticed by the outside world, during that late 1960s and right through the 1970s in Rhodesia (now the independent state of Zimbabwe).

Rhodesian Security Forces faced a mammoth task of combating the continued incursions into Rhodesia of ever increasing numbers of heavily armed terrorists gangs. These gangs were more often than not trained, provisioned and mobilised from large camps positioned within and sanctioned by neighbouring African states sympathetic to their cause.

One method that the Rhodesian High Command employed to combat these incursions was to strike at, and destroy these camps when they were considered too great a threat to national security. During the closing months of 1978, one such strike was planned and executed by aircraft of the Rhodesian Air Force on Westland's Farm, a large training camp on the outskirts of Lusaka, the capital of Zambia which adjoined Rhodesia's northern border.

The strike was led by four Canberra bombers, immediately followed by cannon and rocket strikes from two Hawker Hunters. Synchronised to follow shortly after were four Alouette III helicopter gunships. As the Canberras initiated the attack, two additional Hawker Hunters were positioned over Lusaka airport and a further two were assigned sentry duty over Mumbwa, the Zambian Air Force base. Lusaka was in direct contact with the Zambian Air Force.

Knowing that there were two Hunters in position over Lusaka Airport and two more circling Mumbwa, ready to take up any challenge the Zambian Air Force might make to them, the Squadron Leader commanding the Canberras called up Lusaka International Airport.

"Lusaka Tower, this is Green Leader".

"Uh .. Green .. this is tower," came a puzzled but otherwise unworried reply, after a short pause.

"Lusaka Tower, this is Green Leader. This is a message for the station commander at Mumbwa from the Rhodesian Air Force. We are attacking the terrorist base at Westland's Farm at this time. This attack is against Rhodesian dissidents and not against Zambia or her security forces. We therefore ask you not to intervene or oppose our attack. However, we are orbiting your airfield at this time and are under orders to shoot down any Zambian Air Force aircraft which does not comply with this request and attempts to take off. Did you copy all that?"

"Yes, copied Sir" replied Lusaka Tower.

"Roger .. cheers", said a friendly Green Leader.

Various courteous exchanges took place after this while the attack on nearby Westland's Farm was completed. A Rhodesian Air Force Command Dakota then took over the direction of Zambian airspace from Green Leader, who then moved out.

Meanwhile, the pilot of a Kenya Airlines jet, now considerably delayed in receiving a landing clearance, irritably radioed Lusaka Tower demanding why.

"Who is in control here anyway?"

"Well ..." said the African controller at Lusaka, with dry humour, "I think at the moment the Rhodesians are ..."

~ ~ ~ ~

Having dispatched a Mustang, a member of the tarmac crew watched it taxi out towards the duty runway. He suddenly became aware that the spreader bar, used to restrain the undercarriage legs when the aircraft was being towed, was still in place.

Frantic activity by several persons managed to have the aircraft halted before takeoff. Had the spreader bar remained in place it may have bounced out of its sockets during the takeoff run, with consequences not even to be imagined, or had it remained in place would have prevented retraction of the undercarriage.

The hero of the day was an airman who leapt on a bicycle, chased the aircraft until it stopped, and working far closer than many thought comfortable to a very large spinning propeller, removed the bar and casually pedalled back with the offending item over his shoulder.

Mustang with spreader bar in place

~ ~ ~ ~

Back just before WW2 a young man was in Sydney after learning to fly at the Royal Newcastle Aero Club and heard that an air pageant was to be held at Coonabarabran near his tiny hometown of Baradine in northwest NSW.

He made some phone calls and succeeded in buying a ride to Coona in a Cirrus Moth. On the way the pilot wanted to know exactly where they were so he descended low over a small town they could see, and flew alongside the railway line at about 50ft to read the station name. As soon as they saw "Denman" the aircraft turned, still very low, and flew alongside the main street, where only a few people and one vehicle could be seen. It was in fact a one horse cart with the driver sitting on a backless seat towards the rear of the cart.

As the aircraft was coming towards, and almost over him, he kept his eyes on the Moth by moving his head backwards with the result that he overbalanced into the cart and fell out onto the street, much to their amusement as well as to a bunch of fellows sitting on the kerbside.

The funny thing was that the old carthorse didn't even miss a beat and kept jogging down the street, with the driver in full pursuit on foot after shaking his fists at us. Now, that wouldn't happen nowadays (thank God! Ed).

~ ~ ~ ~

After completing initial flying training on Tiger Moths at Benalla in 1943 a young student was transferred to Uranquinty in New South Wales for conversion to Wirraway advanced trainers.

Soon after arriving at Uranquinty, his instructor, Flt Sgt Williams, was teaching him landings at a satellite drome with repeated circuits and bumps, where several other pupils and their instructors were engaged in the same exercise. Approaching lunch time, and after a short break, it was decided by the instructors that they would return to base taking off and flying in formation to show the "sprogs" how it was done.

On getting into the plane and being informed of this, the student became totally engrossed in the excitement of what was going to happen and completely overlooked fastening both his parachute harness and seat belt. However, on settling down and at the required cruising revs the plane kept dropping back and one of the other instructors indicated that the wheels had not fully retracted and, despite re-engaging the system, continued to do so.

In order to try and force them up the instructor pushed the control column sharply forward and as the rear canopy was open – it was summer time – and not being secured the student was thrown upwards and can vividly recall desperately grasping the side of the plane as he left the seat. On realising what was happening the instructor very quickly pulled the column back and prevented him from being thrown out.

Being in the rear seat the instructor had a good view of what was happening and claimed he briefly saw daylight between the student and the plane which probably was to both frighten the student and teach him a lesson.

CAC Wirraway

~ ~ ~ ~

A quiet Sunday afternoon shift at Derby in Western Australia a few years back is broken by ...

A/C: "Derby this is Qantas nine-one-three on one-two-two decimal one."

DERBY: "Qantas 913, Derby, go ahead."

QF 913: "Yeah ... we can see what looks like a very long runway to the south east of you, wondering if you can tell us what it is and how long it is?"

DERBY: "Qantas 913, it's the new RAAF base, still under construction. The main runway is 3000 metres long."

QF 913: "Ah, good ... we could use that ... we've only got two engines!"

Now when a pilot says something like that, there's a scramble made for the panic button and if you're an FSO or ATC, you start wishing you were anywhere else but at work. But before they all recovered, the pilot came back with ...

QF 913: "It's all OK, we're one of the new 767s!"

~ ~ ~ ~

One common mistake made by crews flying into Melbourne's International Airport is that they sometimes mistake the GA airfield of Essendon for Tullamarine. The 747 we will call 'Atlantic' and the following exchange took place on Melbourne Approach.

ATC: "Atlantic 791, you are six miles to the south east of the field, do you have the airfield in sight?"

A/C: "No ... what's the airport underneath us at the moment?"

ATC: "No, no ... not that one, you're for the one at your two o'clock position!"

The surrounding locals near Essendon would not have been too happy with a 747 on short finals.

~ ~ ~ ~

Back in the '80s a television news reporter and crew were dispatched to spend a week covering major flooding in the border rivers area of northern NSW and southern Queensland. To get around, they were issued with a small aerial fleet of a fixed wing twin and a helicopter.

On the final day of the assignment they flew the JetRanger on an exhausting seven hour journey around the outback, racing to end up at the small NSW bush town of Nyngan to meet an appointment at the local Telecom telephone exchange/repeater station to feed their pictures back to the TV station in Sydney.

Now these were the days when Telecom could have run the show properly if it wasn't for the customers. And if you didn't turn up for your booking within about three minutes of the appointed time well, sorry, bad luck, too late, missed out, queue up and fill in all the forms again!

They arrived overhead Nyngan in the JetRanger with only minutes to spare. But the Telecom compound didn't have a car park, and the local airport was too far away from town by road (anyway, the only taxi was in a town some miles away that day!)

So, they made a command decision; land in the middle of town on a large area of flat, vacant ground on the main street, alongside the railway line. The locals were much in awe as the helo landed, and watched as three likely lads disembarked from the helicopter with various bits of electronic gear under their arms and sprinted to the Telecom exchange (they got there in time!)

The helicopter pilot, a Royal Navy trained, nonchalant, British sort of chap, took all this in his stride, tied down the rotor blades and made himself comfortable in the back with a newspaper.

Comfortable, at least until there was a knock on the window, and Constable Plod announced "You can't park here, mate. You'll have to move along!" The complexities of TV deadlines and the Telecom bureaucracy were a little beyond Plod, who felt the majesty of the law had to be observed, and so proceeded to take the helicopter's "particulars", so a parking infringement notice could be issued!

The locals later assured us this particular constable had previously been stationed "out in the country", and he wasn't quite used to the sophisticated sort of things that happened "in town".

Oh, and the parking infringement notice seemed to get lost in the mail. But those were the days when Australia Post ....

~ ~ ~ ~

One of the challenges of flying the RAAF's Macchi jet trainer was making the trip from east to west of Australia in winter against the prevailing westerlies and being forced to refuel at Woomera (better known as a weapons test range in South Australia) prior to setting off for the 'short hop' to Forrest in WA.

This leg was usually very tight against the jetstream and, if the weather conditions at Forrest were poor, the trip was very interesting indeed. Add to this being in cloud for the entire time in the cruise or complete cloud cover over the Nullarbor, and the trip became one of your more memorable aviation experiences.

The leg distance was just over 400 nautical miles (740km/h) but with a ground speed often reduced to barely 200 knots (370km) from the normal range cruise of 300 knots (556km/h) the leg would be at the maximum of the Macchi's two hour endurance. With only a less than acceptable radio beacon at each end and a temperamental direction finder in the Macchi when flying in cloud with storms and ice, one relied heavily on the accuracy of the met man's forecast winds and then prayed once past the point of no return and out of radio range.

This was the scenario one pilot found himself in – 26,000 feet, in cloud, no sight of ground or water since leaving Woomera, only marginal radio navigation equipment and fast running out of options unless he managed to obtain an accurate groundspeed check.

Hearing an Air Force C-130 Hercules give a position report to Adelaide control, he asked the Herc on frequency for a wind check, to which his reply was:

Defence PR

Aermacchi MB-326

"Alladin 247 – Stallion 10 – winds at flight level 240 and flight level 260 is 260 degrees at 85 knots."

An Ansett 737 also piped up with the news:

"Alladin 247 – Charlie Zulu Echo – the wind up here at FL310 is 280 degrees at 100 knots."

To which his reply was "Thanks for the good news guys".

CZE sarcastically came back with, "Glad we could be of help".

Being fully aware of the 'legal' navigation requirements, Adelaide control got on the airwaves with:

"Alladin 247 – Adelaide control – Are you domestic DME equipped?"

"Alladin 247 – Ah – Negative!"

"Alladin 247 – Adelaide – Are you in cloud?"

"Ah – Alladin 247 – Affirmative."

"Ah – Roger Alladin 247 – best of luck!"

Full of encouragement those procedural control types.

One point five hours and 400nm (740km) late, following a trip in total IMC, Forrest appeared through the 5000ft overcast. Another testimony to heading and airspeed! (Incredible to think that all this is now a thing of the past thanks to something as simple as a $500 hand held GPS.)

~ ~ ~ ~

Some years ago, I had just graduated from my Iroquois conversion course with the RAAF and was sent off on an exercise in northern NSW.

The squadron's job was to support the hostile air threat with Winjeels providing Forward Air Control (FAC) for a squadron of 'baby knucks' flying Macchis from 2 0CU. The Macchis were well suited to the hilly terrain we were operating in, as the Mirage would have had trouble turning in such tight areas without crossing into Queensland halfway through a turn!

During one of the troop lifts on which I was copilot, we were returning from dropping off another 'chalk' of troops when we were bounced by a couple Macchis. We had the frequency that the Forward Air Control (FAC) Winjeel was using to guide the Macchis and we were expecting them, so it was "all eyes out" so we would know when we had been spotted.

Anyway, the loadmaster on the left side, behind him, was the first to spot them and called that they were "rolling in". Suddenly, he yelled in a voice which had rapidly shifted up a couple of octaves "They're shooting at us, they're really shooting at us!!!" He looked around and saw the lead Macchi head-on and approaching us from the side. "It's OK mate", I said, "that's the anticol you're looking at!"

The 'loady' had seen the Macchi's belly anti collision light. Because we were looking at the Macchi head on, as the light rotated, it appeared to flash at us.

We didn't hear much from the 'loady' for the rest of the flight.

A second incident that day came when we heard the Macchis rolling in on us again after they had shaken the previous pair off. We heard one call to another that he had spotted a lone Iroquois in a valley and we knew it was us, so we started to try to out manoeuvre them. Because of their turning circles, both Macchis overshot and had to turn around in the next valley before coming back to try again.

Imagine their frustration when they couldn't find the Iroquois. Had they looked at the small copse of trees in the centre of the very large paddock they'd been caught in, they would have noticed a lot of movement from the trees on an otherwise still day. Yep, we were hovering and hiding in the middle. It worked too!

~ ~ ~ ~

As a newly qualified civil pilot at the Victorian general aviation airfield of Moorabbin, one of the first things our hero did was grab a lady friend he was trying to impress, and casually ask if she'd like to come flying – with himself as the pilot, of course. She accepted, and after suitably impressing her with his new found status, they returned from the Moorabbin training area.

Given instructions to set up for right base on runway 13R, the pilot complied. Turning final, he found himself next to a RAAF Caribou on finals for runway 13C, a grass strip, where it was conducting STOL demos.

Now you must remember that this was his first flight as a licenced pilot, it was only the second time he had set up for 13R (usually having landed on 17L), and was right next to the (relatively) huge Caribou. Feeling decidedly uncomfortable, and sure that discretion was the better part of valour, he calmly picked up the mike, and adding power, announced:

"November Alpha Kilo, going round."

From out of the ether came an unidentified voice ....

"Chicken."

~ ~ ~ ~

The Seattle bound Delta 767 was very much full as it departed LAX.

Most passengers noticed that there was a degree of tension between a female and male flight attendants. There were constant glances between the two, and some seemed to verge on deep concern. About an hour into the flight the two finally met in one of the aisles, with some degree of obvious distress and quickly ventured off to their duties once more.

Approaching the end of the flight, the captain announced that they had commenced descent and to fasten seatbelts. Then an uncertain voice came over the PA system moments later and managed to mumble the words "Joanne, I love you, and I would be the happiest man alive if you would be my wife". Not the most original of proposals, yet 'Joanne' was so overjoyed that she left the rear galley and came bustling up the aisle, crying, to greet him.

Others in the cabin were astounded, and their embrace was applauded and cheered with such volition that passengers without sight of it forgot the 'fasten seatbelt' sign and poked their heads from every corner to catch a glimpse of what the commotion was all about.

At the end of the flight the newly engaged couple were responsible for the 'goodbyes' at the exit door and an enormous queue formed to congratulate them. It is certainly a happening that does not occur on every flight!

~ ~ ~ ~

Heard on board an Eastwest Airlines flight from Sydney to the Gold Coast just prior to takeoff, a flight attendant asked over the PA system ... "Is there a passenger Short on the aircraft? And if so please press the call button."

A "wag" passenger sitting in front of him duly pressed the call button, and when the hostess arrived said to her ... "Well, I'm really only four feet eleven, and not only that ... I'm broke, so what can I do for you my dear?"

The surrounding passengers fell about in laughter, leaving a rather embarrassed looking flight attendant wondering what the hell was going on!

Eventually, the real "Mr Short" made himself known and even he was embarrassed by the situation.

~ ~ ~ ~

This exchange was heard one evening on Adelaide Approach frequency. The callsigns have been changed to protect the guilty.

VH-TBZ (Australian 727): "If we can track direct to the omni for a visual approach, we could pass the 737".

APPROACH: "I admire your optimism!"

VH-CZZ (Ansett 737): (Said in very sarcastic tone) "Try again fellas!"

The following comment had no callsign attached, but the suspect would have to have been the slightly peeved 727 Captain: "Why don't you just wind that Tupperware Twin up!"

~ ~ ~ ~

On the radio one day in the early 1950s, Adelaide's Parafield Tower was talking to a TAA Convair, the so-called Cannonball Service from Sydney. It was the hot ship of the times, and Parafield Tower advised the Convair crew of traffic, being a DC-4, and two DC-3s, plus two Tiger Moths in the circuit, and three Tiger Moths on local flying. After being so advised, the Convair crew member on the radio replied ... "

"Okay Parafield, we've got our butterfly net out!"

~ ~ ~ ~

The first Australian Airlines 737 to be painted up in the striking Qantas "flying kangaroo" livery (VH-TAK) was taxying to the domestic terminal at Adelaide International Airport.

GROUND: "Have they called you Joey yet?"

(No response from the aircraft)

GROUND: "Have they called you Joey, the junior kangaroo?"

VH-TAK: (chuckling, then an American drawl) "Not yet mate ... you're the first!"

Qantas 737 VH-TAK

Bill Lines

~ ~ ~ ~

This classic exchange was heard at the time of the recent 6.8 Richter Scale earthquake in Los Angeles, at LAX. This air-ground conversation took place during one of the aftershocks felt across most of California. This particular aftershock brought down ceiling panels in the terminal building where our contributor was seated.

LAX TWR: "Attention all aircraft, there is a strong earthquake in progress!"

UNITED: "United 72, that's good man ... I need all the excuses I can get when I'm landing!"

~ ~ ~ ~

Late evening, Sydney Kingsford Smith Airport, many years ago, with two Ansett Fokker Friendships being vectored east of the coast for a Runway 16 arrival.

APPROACH: "Foxtrot Charlie Charlie, cleared for final make visual approach Runway 16, contact Tower when established."

VH-FCC: "Foxtrot Charlie Charlie."

APPROACH: "Foxtrot Charlie Delta, track east of the coast for a right circuit Runway 16, you are number two to a company Friendship, turning final."

VH-FCD: "Foxtrot Charlie Delta."

VH-FCC: "Foxtrot Charlie Charlie established on final, Captain requests how close we were to other aircraft" (sounding stunned).

APPROACH: "What other aircraft?"

VH-FCC: "Below us, when we turned final!"

APPROACH: "Cripes, I've got something on radar now. Thanks. Contact Tower 120.5, goodnight."

VH-FCC: "Goodnight and good luck!"

APPROACH: "Foxtrot Charlie Delta, vector heading for unknown traffic your twelve o'clock, please advise if sighted."

VH-FCD: "Yeah, confirm traffic our twelve o'clock low now, lights on ... "

APPROACH: "Err, Foxtrot Charlie Delta, continue on that heading and advise!"

(Some minutes later ...)

VH-FCD: "Jeeze, it's all lit up like a bloody ship. Cripes, it's a great big bloody passenger ship!"

~ ~ ~ ~

A Qantas aircraft was heard requesting the latest football scores. Sydney Control obliged and passed the scores to the aircrew, and shortly after, a commuter airliner asked for a repeat because he hadn't copied the scores when they were read out. Sydney Control once again read out the scores for the second aircraft.

Then, would you believe it, another aircraft wanted them as well! This was Sydney Control's response.

"Listen folks ... there are some people here at ATC who do not want to know the scores till they see the game replayed on TV, and I tell you now ... if I repeat 'em again, I'm seriously going to have my head bashed in!"

~ ~ ~ ~

At Townsville, the RAAF uses two BAK12/14 hook cable systems on the main Runway 01/19 about 600m from each end of the runway. They are referred to on air as the "northern hook cable" and the "southern hook cable". The system is capable of being used by the F/A-18 and F-111. The maintenance crew for this system is called "Tennis Team" on air, which came about from the previous arrestor system in use, the BEFAB 21.2 Arrestor Net. This structure, when fully raised up, looked like an outsize tennis net.

A couple of years ago, the "tennis team" was on the runway, having just completed an unscheduled maintenance of the hook cable system. The following transmissions took place on Ground frequency:

TENNIS TEAM: 'Ground, Tennis Team, we have completed maintenance to the southern hook cable, and we're now vacating the runway. Full length is available."

FEMALE ATC: "Tennis Team ... just the way I like it!"

~ ~ ~ ~

In this story, in the early days following the Qantas/Australian Airlines merger, the callsign of the Qantas 767 was "Oscar Golf Delta", not the usual "Qantas ... " with the flight number attached. The pilot is still not fully resigned to having to revert from the Qantas callsign back to registration letters on domestic flights. And on this occasion he got the letters slightly muddled!"

MEL ATC: "Oscar Golf Delta, change to Melbourne Control now."

VH-OGD: "Golf Oscar Delta (GOD) ... what was that frequency?"

MEL ATC: "With a callsign like that Sir, you should already know the next one to dial up!"

The quick wit of ATC left 'GOD' lost for words.

~ ~ ~ ~

From the flightdeck of a Dove, bound from Hail to Riyadh, in Saudi Arabia, comes this tale.

RIYADH TWR: "Hotel X-Ray, you are second in line after the Boeing 737 on your left."

The pilots both searched the sky out to port, but all that could be seen was a group of Strikemasters zipping off to their training area and a DC-8 taking off. No 737!

The Dove still had fifteen miles to run and they were not yet on finals, so the crew just carried on with vigilance. One of the pilots turned to look at the starboard engine ... and you guessed it ... there it was, about five miles away and overtaking them ... a Boeing 737 of Saudia Airlines.

The penny finally dropped. The ATC man was viewing the situation from his position and not allowing for the Dove's rather different outlook. If the crew of the Dove put themselves mentally in the ATC position, it surely would have looked like the 737 was off to the port side! Oh well, definitely a case of watching your 'six o'clock' when in foreign airspace!

BAe

Hawker Siddeley Dove

~ ~ ~ ~

This encounter was overheard at Sydney Airport one recent Sunday.

Speedbird 11 had been holding for some time at taxiway Foxtrot, for a Runway 16 departure.

TOWER: "Speedbird 11, Cessna on short final, line up behind that aircraft."

SPEEDBIRD: "You want us to line up behind the microlight! Okay."

The Cessna then seemed to take ages to land and roll clear of the runway.

SPEEDBIRD: "All four hundred and two of us on board the Speedbird one-one are enormously impressed by that dreadful delay, Tower!"

~ ~ ~ ~

Several years ago Bruce was offered the chance to deadhead on an overnight freight run in a Douglas DC-3 from Melbourne to Canberra, then Sydney, and return.

Some hours into the return legs, with all of the cargo removed from the aircraft, Bruce was offered the chance to sit in the right seat, don a headset, and be shown through the flight controls. The First Officer was on the left, and gave a very full and fascinating description of what each instrument was and what function it performed. The description had continued for some period of time, when Bruce noticed that the microphone switch belonging to the F/O was on "radio" and not "intercom". Bruce wanted to know if the switch was in the correct position, and keyed his mike to ask the First Officer. A voice on the radio emanating from the ground station interjected with:

"No ... it's not!"

A look of total dismay on the F/O's face, and a hasty repositioning of the switch led Bruce to believe that they had been overheard on the ground throughout the lengthy demonstration of the controls and instruments!

~ ~ ~ ~

The air traffic controllers in the far north of Australia can be quite laid back and they often have a zany sense of humour. On one trip, an Ansett F28 departure ex Cairns went something like this:

VH-FKJ: "Tower, Foxtrot Kilo Juliet, request line up."

TOWER: "Foxtrot Kilo Juliet, go ahead, line up."

VH-FKJ: "Tower, Foxtrot Kilo Juliet, ready."

TOWER: "Foxtrot Kilo Juliet, traffic is two ducks crossing the main runway complex at slow speed, south-east bound ... They're coming back. So they're clear of your path, and you're clear for takeoff!"

~ ~ ~ ~

This story concerns an Australian Airlines 737-400 during one of the refuellers' stoppages that took place during 1992.

When refuellers in Brisbane were out, it was not uncommon for various aircraft to land at Maroochydore (on Queensland's Sunshine Coast north of Brisbane) and take on fuel for their destination after Brisbane. This particular evening, VH-TJG had refuelled at Maroochy and was about to depart for Melbourne. Maroochy Tower cleared it for takeoff, and then instructed it to contact Brisbane Control with departure details.

The conversation with Brisbane went something like this:

VH-TJG: "Brisbane, this is Tango Juliet Golf, departed at time three six, and we're currently six miles south of Coolangatta."

At this point in time, the aeroplane was passing overhead Mooloolaba, some 200 kilometres north of Coolangatta.

BNE CONTROL: "Tango Juliet Golf, this is Brisbane Control, squawk code one four three two."

The transmission was acknowledged, then several minutes later ...

VH-TJG: "Brisbane Control, Tango Juliet Golf, correction to those details, we are actually just south of Maroochy. I think I said Coolangatta before!"

BNE CONTROL: "Tango Juliet Golf, we knew what you meant, and we most certainly know where you are!"

VH-TJG had been going through a bad time that night, having already been to Maroochy before Brisbane, and had taken on fuel for Sydney. Apparently, in the shuffling of aircraft to overcome some unknown problem, it was altered to a Melbourne flight, and of course needed more fuel than it had taken on board. So it was back to Maroochy to top up!

Such aircraft reshuffles have been known to cause problems. Problems such as happened to Ansett and EastWest one time, when an Ansett Express F28 went unserviceable. Ansett Express grabbed an EastWest BAe 146-300 to service its route to Coffs Harbour, but this in turn left EastWest short for the Maroochy service, so Ansett provided a 737.

It was not until the aircraft was halfway there (or so the story goes!) that it was realised that Ansett/EastWest had no 737 steps for passengers at Maroochy. BAe 146 steps leave quite a large jump from a 737! Being a common user terminal, however, the staff of all the various airlines at Maroochy were on very friendly terms with each other, and often helped each other out, knowing the favour would be returned when the "shoe wound up on the other foot".

At the last minute, the other airline ... Australian, provided the use of its last spare set of steps!

So this was the final outcome:

1. Passengers booked on EastWest!
2. They flew Ansett!
3. But they finished their trip on Australian Airlines steps!

~ ~ ~ ~

Solomon 700, a leased Boeing 737-400 (from Qantas), talking to Qantas Brisbane.

SOLOMON 700: "Qantas Brisbane, Solomon 700."

QF BRISBANE: "Good afternoon Solomon 700, this is Qantas Brisbane, go ahead."

SOLOMON 700: "How are you dear Sir? Blocks on the hour, nil sickness on board, registration is Hotel Four Sierra Oscar Lima, we have eight crew disembarking and no special requirements."

QF BRISBANE: "Copied that Solomon 700. It's Bay 1 for you today!"

SOLOMON 700: "Goodness, we ARE important aren't we. Imagine ... Bay 1!"

QF BRISBANE: "Don't get too excited there, Bay 1 is the closest to the Qantas Domestic Terminal!"

SOLOMON 700: "Oh dear me ... just delusions of grandeur again!"

QF BRISBANE: "See ... you're not THAT special after all!"

It was great to hear two people, obviously professionals, enjoying their work.

~ ~ ~ ~

VH-BKS: "Essendon Tower, helicopter Bravo-Kilo-Sierra for the northern pad."
TOWER: "Bravo-Kilo-Sierra, track for the northern helipad, and what type are you?"
VH-BKS: "Kawasaki BK 117 Sir."
TOWER: "I've never seen one of those before!"
VH-BKS: Well before I land, I'll fly past the Tower so you can have a good look if you want."
TOWER: "Affirmative, fly past the Tower, that's approved ... oh yes!"
(And then ... when VH-BKS is about to land ...)
TOWER: "Wonderful ... just wonderful."
VH-BKS: "Yes it is, isn't it?"
TOWER: "I love that colour scheme."
VH-BKS: "Yes, it's good, that's for sure!"

~ ~ ~ ~

Two guys were flying themselves on a local trip near Moorabbin Airport. The day was pretty terrible with a 40 knot (74km/h) wind from the north, and all operations were taking place from runways 35 Left and 35 Right. As they held their light aircraft at the holding point and swapped from Ground to Tower frequency on 118.1 MHz, they overheard the following encounter between the Tower and Mike Foxtrot Yankee, a Piper Warrior.
TOWER: "Mike Foxtrot Yankee, it would be really nice if you could make it down."
VH-MFY: "Look ... we're trying ... really!" Due to the very strong wind, VH-MFT was having a whole heap of trouble attempting to land. Some time later, after they had become airborne in their aircraft, Mike Foxtrot Yankee finally had made it down.
VH-MFY: "Hey ... we made it!"
TOWER: "Congratulations, we're very happy for you."

~ ~ ~ ~

A pair of Cessna Skyhawks inbound in formation from the Bankstown training area to the west of the field, with the unlikely radio callsign of "Banana Formation"!
Yes, you guessed it, you really wouldn't expect a callsign like that to go uncommented on, by either controllers or other pilots, now would you?
Well, "Banana 1" and "Banana 2" (and they weren't even yellow!) were cleared to land by Bankstown Tower.
Then ...
UNKNOWN AIRCRAFT: "Are you in pyjamas?"
BANKSTOWN TWR: "And I thought we were eating mandarins!"

~ ~ ~ ~

Some routine traffic between Melbourne Arrivals Control and an inbound United Airlines Boeing 747. The aircraft said something akin to ... "Maintaining 300 knots until touchdown".
MELBOURNE APPROACH: "Ah ... United, confirm you really meant maintain 300 knots until 20 miles from touchdown Sir?"
UNITED 747: "Yeah, yeah, I goofed ... however I must say these things have carbon brakes, and they really are very good."
MELBOURNE APPROACH: 'Yes, well it would have been real good to watch!"

~ ~ ~ ~

CESSNA: "Tower, Cessna November Uniform Tango, I'm out of fuel!"
TOWER: "Roger November Uniform Tango, what's your position, and have you got the airfield in sight?"
CESSNA: "Oh yeah ... I'm actually on the south ramp, and I just want to know where the BP fuel truck is!"

~ ~ ~ ~

Bankstown Tower, western Sydney. One lone Piper Tomahawk is doing touch-and-go's. There is no other aircraft in the circuit, and it's nearing last light. The Tomahawk has done a lot of touch-and-goes, and the controllers are clearly getting quite giddy watching them.
TOWER: "How many circuits are you planning to make?"
TOMAHAWK: "We'd like to do two or three more."
TOWER: "Well, we were just working out your landing fees, and your bill is in excess of thirteen thousand dollars!"
A long pause ... and then ...
TOMAHAWK: "Are you kidding? That was our last one!"
TOWER: (laughs) "Yeh, just kidding."

Piper Tomahawk

~ ~ ~ ~

Late one night a contributor was flying near central Washington when he overheard the following conversation on the radio:
AIRCRAFT: "Seattle Centre, Mickey Mouse at one-nine-oh!"
ATC CENTRE: "Hi ya Mickey ... squawk ident. Who ya got on board tonight?"
AIRCRAFT: "Oh let's see ... we got Mickey, Donald Duck, Goofy, Snow White, and some of the Seven Dwarfs with us this time!"
The two pilots monitoring the call looked at each other in surprise with the comment made "Must be some silly doofus fooling around on the radio!"
But Mickey Mouse was for real! The aircraft was a Gulfstream bizjet registered N234MM, Walt Disney Corporation's Gulfstream I, on the way back to California after a Disney show in Seattle! And you guessed it, they *really* had Mickey, Goofy and the whole gang on board.

~ ~ ~ ~

Now over to the UK with a conversation between the tower and a freighter '676'.

TOWER: "National 676, cleared for takeoff, report passing two thousand."

NATIONAL 676: "676 copy, we'll report passing two thousand."

Shortly afterwards ...

NATIONAL 676: "Southend, National 676 is passing two thousand on climb to two-five-zero."

TOWER: "Thank you, call London Control 128.6.

Soon after the freighter lost a door ... yes, that's right ... a door!

NATIONAL 676: "Mayday Mayday Mayday, London Control, this is National 676, four miles west of Southend, two thousand five hundred. I've lost the door, and am returning. We're climbing to four thousand, and returning to Southend."

LONDON CTL: "National 676, London Control, are you in control of the aircraft?"

NATIONAL 676: "No more than usual my friend!"

~ ~ ~ ~

This was one of the last Continental Airlines' transmissions monitored from "Continental one-six" before the airline withdrew its services from Australia.

CONTINENTAL 16: "Operations, one-six."

CONTINENTAL COMPANY: "Go ahead Continental sixteen."

CONTINENTAL 16: "Off zero-six two-three, slash four-zero, Guam one-five five two."

CONTINENTAL COMPANY: "Hey, aren't you going to Honolulu?"

CONTINENTAL 16: "Yeah, but first we were thinking we might stop off in Guam and get something decent to eat!"

CONTINENTAL COMPANY: (laughing) "Have a good trip guys!"

~ ~ ~ ~

An Ansett 737 was following a Garuda Indonesia A300 into Adelaide which had just been handed over to the Tower by the Approach Controller. The Controller then spoke to the Ansett Boeing 737.

APPROACH: "Charlie Zulu Mike, come back to approach speed now. You are following an Indonesian Airbus on a four mile final. He is very slow!"

VH-CZM: "He's slow all right. What's his exact speed?"

APPROACH: "One hundred and ten knots. Frankly, I don't know how he's staying in the air!"

VH-CZM: "Yeah ... we thought it was a boat!"

~ ~ ~ ~

A Beech Duchess from the Australian Aviation Academy was doing a practice instrument approach to Adelaide. All the instructions from the Controller had been acknowledged by a youthful sounding male student, until all of a sudden, the following instruction was acknowledged by the female instructor.

APPROACH: "Gee, your voice sure has broken!"

INSTRUCTOR: "Yep, I've had the big operation!"

~ ~ ~ ~

The next item involves aerial operations in Arnhem Land, in the Northern Territory. Most exchanges with local Flight Service are pretty relaxed, because the pilots and the FSOs all know each other and also know their aircraft well. Occasionally, when the pilots fly to the Big Smoke (Darwin), they have to contend with military Air Traffic Controllers, who are always on the ball. So they try to behave, and be alert and professional on the radio! Here is an exchange from just such a visit to Big City controlled airspace!

VH-PGJ: "Darwin Control, good morning, Papa Golf Juliet, 75 miles inbound on the 083 degree omni radial, maintaining six thousand five hundred. Request Airways Clearance please."

DARWIN CTL: "Papa Golf Juliet, contact Darwin Control 129.2."

VH-PGJ: "Papa Golf Juliet" (thinking he had called on the wrong frequency). After repeating the request call on 129.2, Darwin Control responded with his clearance. Wondering if he had definitely made a mistake with the previous call, he then asked for a clarification ...

VH-PGJ: "Ahhh ... Control ... how do I know when to call you either on 123.8 or 129.2?"

DARWIN CTL: "Papa Golf Juliet, you call Darwin Control on 123.8 after 7 o'clock!"

VH-PGJ: "But it's already 7.30."

DARWIN CTL: "Yeah ... well you see ... they're running a tad late!"

~ ~ ~ ~

During the 1950s No 24 RAAF (City of Adelaide) Fighter Squadron was one of several Citizen Air Force 'City' squadrons formed to train civilian pilots and ground staff.

In May 1956, 24 Squadron deployed to Townsville in far north Queensland for operation Comax Sprat, a training exercise combining RAAF bomber and fighter squadrons and involving both citizen and permanent Air Force personnel. At that time, 24 Squadron was equipped with CAC Mustang fighters and during the exercise it was necessary to move the squadron's aircraft from Townsville to Cairns.

All went according to plan until one Mustang piloted by a young recently graduated civilian pilot contacted the ground rather hard while landing at Cairns airport. Believing that damage to the undercarriage may have occurred, the pilot wisely decided to abort the landing and go around again so that the extent of any damage could be assessed from the ground.

On retracting his undercarriage, it was discovered that only one wheel came up. The port undercarriage strut had apparently fractured as a result of the impact and the wheel dangled uselessly under the aircraft.

After some deliberation, it was decided to execute a wheels up landing (or wheel up in this case), and after all emergency services were in position, the pilot ejected his cockpit canopy and lined up on the grass verge of the runway. For one so inexperienced he completed a textbook wheels up landing amid clouds of dust and grass and finally came to rest elated that he and his aircraft were still in one piece.

Imagine his amazement, therefore, when he looked up and the first thing he saw through the dust haze was a burly Queensland policeman, one big boot planted firmly on the wing, and with a notebook in hand asking "Now sir, could I please have your name, address and driver's licence please?"

~ ~ ~ ~

Monitored on Sydney Control between control and Cathay Pacific flight 101.

CX 101: "Ah ... Control, we've got a proposition for you. At 130 miles, if we could make one large orbit, that'd put us in the correct position to start our approach for a landing at time one-eight." (Followed by aircrew laughter)

ATC (American Controller – and a damned good one): "OK Cathay, this is what I got for ya, you got the first part of your request, but after you do that, I'm not sure you're going to be able to keep on coming in!"

CX 101: "Oh ... why?"

ATC: "Well Cathay, you gotta understand ... you're dealing with a pion here and not somebody capable of making those kinda decisions!"

~ ~ ~ ~

Scene: Britannia charter Munich to Luton, a Boeing 737. It reaches its cruising flight level and is given this advice by German ATC.

CONTROL: " ... be advised that a Caledonian BAC One-Eleven is eight miles behind, enroute Bulgaria to Gatwick, and also you can be on the lookout for two Luftwaffe F-104 fighters, crossing left to right. Please report sighting and passing behind the crossing fighter aircraft."

Suddenly, across the ether came the familiar World War 2 fighting words ... "T-A-L-L-Y H-O!" (Both Britannia and British Caledonian then in turn pleaded innocent to all charges!)

British Caledonian BAC One-Eleven

Rob Finlayson

~ ~ ~ ~

One of the casualties of the US air traffic controllers' strike and mass sackings was correct voice procedures. This incident illustrates.

It was a 40 degrees Celsius scorcher at Washington National. United 41, a heavily laden early model Boeing 727, was cleared for a right turn departure with noise abatement over the Capitol. When the pilot held the aircraft for a few seconds while power built up, the controller, complete with a southern drawl, called out ... "United 41, Go! Go! and don't you all forget to noise abate!"

After struggling for altitude in the hot still air over Washington without reducing thrust even for a fraction of a second for noise abatement, the pilot called the Tower. He said in a measured and educated voice ... "United 41, off at three five. Now Tower, come back to me with everything you said after Go, go, go?"

~ ~ ~ ~

Some Mooney aircraft are a little difficult to bring back to a good landing speed, a situation illustrated at Canberra Airport recently, when instead of coming over the fence at a good ol' 75kt (139km/h), we saw what must have been 90kt (167km/h) over the piano keys! The result ... bounce ... bounce ... bounce, six times already! In sheer frustration, as the Tower men watched in amazement, the hapless pilot increased power, flared again then finally set the aircraft down on touchdown number seven. There was stony silence. Then the Tower Controller came to life. "Mike Oscar Oscar ... as far as I'm aware, you were cleared for only one landing Sir, not seven!"

A Mooney 201

~ ~ ~ ~

And now a piece which occurred on a flight from Palm Springs to Las Vegas in a Cessna Conquest, registered something like N141NS.

The pilot called Las Vegas Approach as he began his descent. "Las Vegas, 141 November Sierra is with you, overhead Hoover Dam, leaving 15,000 for 12."

The controller sounded hassled when he replied. "Hold it right there 41 November, there's a biggie comin' up under ya!"

The perplexed pilot asked ... "Do you want me to stop the airplane right here?"

A new voice from the Tower replied ... "141 November Sierra, maintain 15,000 and make a right orbit, five minutes. Traffic is departing heavy and ... err ... we're training some new folks down here. Get it?"

~ ~ ~ ~

A conversation ensued between a female crew member and the operations controller. It went something like this ...

OPERATIONS CONTROL: "Err ... could you pass on to the Captain that a very important passenger has just boarded the aircraft."

AIRCRAFT: "OK ... I'll get the girls to bring him up front."

OPERATIONS CONTROL: "Yeah ... actually, we've been perving at *her* for the past 20 minutes!"

AIRCRAFT (male this time, obviously the Captain): "What was she wearing?"

OPERATIONS CONTROL: "Not very much!"

~ ~ ~ ~

The following exchange was overheard late one night at Perth Airport.

AIRCRAFT: "Perth Tower ... evening to you Sir, Silverbird two-four is on final for Runway two-one."

PERTH TWR: "Silverbird 24, good evening, runway 21 wind calm, QNH 1008, report when aligned on the VASI."

A short pause followed, then back came the completely unruffled English voice from the Boeing 747 ...

AIRCRAFT: "Silverbird 24 ... umm, er, by the way, if it's not too much trouble, could you turn the VASI on please?"

~ ~ ~ ~

The story goes ... Coolangatta Airport, with a CFI (Chief Flying Instructor) waiting his turn to enter the runway while a domestic Fokker Fellowship was on final. The CFI completed his pre-takeoff checks and looked up to see the F28 make a classic copybook touchdown, an absolute greaser. A voice then crackled over the radio, emanating from a Beech Baron in front of the CFI. "Very nice landing Sir!" said the guy in the Baron. The jet rumbled to a full stop, then began a turn to backtrack on the runway to the nearest taxiway. All of a sudden, the F28 Captain accepted the adulation, took a 'bow' and then skited ... "You should see it when I have my eyes open boys!"

~ ~ ~ ~

The following conversation took place between Sydney Control, East-West 165, an F28 jet, and Qantas 17, a Boeing 747 Combi.

East-West 165 was inbound to Sydney ex Norfolk Island, while Qantas 17 was outbound ex Sydney for Honolulu and Los Angeles. Only one thousand feet of vertical separation existed, with East-West 165 at FL280 and Qantas 17 at FL270. Qantas was being made to hold FL270 before climbing to FL330, due to the opposite direction traffic F28 jet. Once the aircraft visually sighted and passed each other, Control would allow Qantas to climb higher.

EW 165: "Ahh ... Qantas 17, East-West 165, we have you sighted out to our right and below us."

QF 17: "Ahh ... sir ... we have no contact."

EW 165: "Look out to your right ... about NOW!"

QF 17 (American accent): "Still no contact Sir!"

EW 165: "Well we have you ... and we're passing now."

QF 17: "Can't see you Sir, we're in and out of broken cloud."

EW 165: "Well we got you."

QF 17: "Ahh, yeah ... see you now ... the Qantas one seven confirms one six zero DME Sydney, sighting and passing opposite direction traffic F28, East-West 165."

CONTROL: "Thank you both!"

~ ~ ~ ~

This one is from Papua New Guinea, of a situation which occurred many years ago.

The morning mists are finally clearing from the Highlands and aircraft are busy doing their runups and calling in to the Flight Service Unit (FSU) with their details. However, a very proper executive aircraft pilot is already on the airwaves from an outlying strip, not only filing a lengthy flightplan, but requesting enroute winds, notams and who knows what else ... on and on it goes! Finally, a laconic Australian voice breaks in: "Yer three minutes are up ... are you extending?"

~ ~ ~ ~

Heard one evening was an international flight calling Honolulu Radio on frequency 5643 Khz at 1240 zulu, to which Honolulu replied and then asked was the aircraft inbound to Honolulu, or on ground at Honolulu. The HNL radio-op was clearly set down on his posterior when the 747 crewman informed him he was 'on the ground Sydney'. Nice one!

~ ~ ~ ~

This story relates to a hot air balloon ride over Canberra, one of eight flying that day from Balloons Aloft.

Takeoff from the National Library lawns at dawn was led by ex-world champion Peter Vizzard. One of the balloons, a large 10 passenger unit known as the 'cattle truck' was piloted by a Frenchman, who'd only been in Australia three weeks. The pilot explained to his passengers that the Frenchie had been 'warned' many times about carefully choosing zones for landing sites clear of trees, because of the 'current plague of drop bears'. The flight over Lake Burley Griffin including some very low level manoeuvres, with the fun including allowing the basket to skim the waters of the lake. Large carp were subsequently seen near the murky water's surface. One of those to observe these giants going about their business was 'le pilot' who quickly came up on his radio transceiver from the 'cattle truck' balloon. At the time he was skimming the water. "Zar are very large vish in zee vater, could zey be zharks?"

Hot air balloon over Canberra

Ian Hewitt

There was no reply, with seven other pilots and some passengers holding back huge smiles. Moments later, the serenity of the scene was blasted to bits with the roar of propane burners rocketing the 'cattle truck' skywards!

It should be mentioned here that as Canberra is some 150km inland, the lake is freshwater.

~ ~ ~ ~

Qantas 44 returning to Sydney from Auckland talking with Qantas Operations.

QANTAS 44: "Qantas Operations, one of our flight attendants has locked his bar float inside his bar trolley. Can you have someone from the bond area meet the aeroplane to unlock the trolley and allow him to recover the float? They can then lock it up again".

QANTAS OPS: "Qantas 44, you have two chances of getting them to do that. None, and Buckleys!"

QANTAS 44: "Okay, I'll break the good news to him!"

~ ~ ~ ~

A light aircraft is transitting a busy US Control Zone. As many readers will appreciate, US terminology is somewhat different to that found in Australia.
APPROACH: "November eight-six charlie, say altitude."
A/C: "Altitude?" (obviously in a frivolous mood)
APPROACH: "November eight-six charlie, say altitude."
A/C: "Altitude?"
APPROACH: "Uh huh ... November eight-six charlie, now say 'Cancel IFR!"

~ ~ ~ ~

This incident involves an F28 which has just landed on Runway 21 at Perth, taxying behind a 747 which was due to depart.
F28: "Is that you Jimbo?"
747: "Yeah ... hi Monty!"
F28: "Dinner on Friday night fixed up with you two?"
747: "Sure ... no problems ... see you then!"
F28: "Okay Jimbo ... safe flight."
(A few seconds silence was heard to follow ...) then,
F28: "Oh ... what a B-I-G red tail!"
747 (different voice): "All the better to T-H-U-M-P you with my dear!"
F28: "We're gone!"

~ ~ ~ ~

The following conversation was heard on a Townsville Control frequency one evening.
QANTAS 25: "Control, Qantas 25, are you busy?"
TOWNSVILLE CTL: "Qantas 25, negative."
QANTAS 25: "Could you ring Merle's Restaurant in Townsville, and ask for Julie, and tell her Bill was just flying over, and he says 'hello' to her."
TOWNSVILLE CTL: "Qantas 25, standby."
Several minutes later.
TOWNSVILLE CTL: "Qantas 25, message passed. Julie says 'hello', and I have a telephone number for you to ring next time you're in town, when you're ready to copy."
QANTAS 25: "Go ahead."
TOWNSVILLE CTL: "The number is 871 234."
QANTAS 25: "Qantas 25 ... err, thanks for that, and I trust you will destroy that number for me."
TOWNSVILLE CTL: "Qantas 25 affirm."
UNKNOWN AIRCRAFT: "Yeah ... he might destroy it, but I most certainly won't, ha ha!"

~ ~ ~ ~

A domestic widebody had just landed on Runway 14 at Coolangatta and was back-tracking. A Piper Comanche was on the taxiway heading towards the runway. The Piper turned on the runway to be greeted by ...
TOWER: "Papa India Papa ... hold on, there's a rather large impediment on the runway."
(No ID – but obviously from the widebody) "Excuse me!"
TOWER: "We're all a bit concerned about our weight lately!"
(No ID – but again the widebody) "Want to have a go at chicken pal?"

~ ~ ~ ~

Hornets 'Maple Hawkeye 1, 2 and 3' were tracking indian file at flight level 290, northbound, and when nearing Tamworth Control Zone, all in turn called Sydney Control requesting descent to 250ft AGL and 'no-comms' for a low-jet-route high speed flight.

The present time when 'Maple Hawkeye 3' called up was 0113 Zulu. When quizzed about his 'no-comms' by the American controller at Sydney ATC, 'Maple Hawkeye 3' gave "0652", very obviously an incorrect figure. It just didn't make sense. The following conversation then began:

SYDNEY CONTROL (AMERICAN): "Maple Hawkeye 3, check your descent and no-comms time will you. Time is now 0113. You gave descent and no-comms estimate of 0652."

MAPLE HAWKEYE 3: "Uh ... standby."

UNKNOWN AIRCRAFT: "Are they F/A-18s?"

SYDNEY CONTROL: "Yeah."

UNKNOWN AIRCRAFT: "I think he's gone away to re-stack his blocks!"

~ ~ ~ ~

The dazed crew of a Japanese trawler was plucked out of the Sea of Japan clinging to the wreckage of their sunken ship. Their rescue, however, was followed by immediate imprisonment once authorities questioned the sailors on their ship's loss. To a man they claimed that a cow, falling out of a clear blue sky, had struck the trawler amidships, shattering its hull and sinking the vessel within minutes.

They remained in prison for several weeks, until the Russian Air Force reluctantly informed Japanese authorities that the crew of one of its cargo planes had apparently stolen a cow wandering at the edge of a Siberian airfield, forced the cow into the plane's hold and hastily taken off for home.

Unprepared for live cargo, the Russian crew was ill-equipped to manage a now rampaging cow within the aircraft's hold. To save the aircraft and themselves, they shoved the animal out of the cargo hold as they crossed the Sea of Japan at an altitude of 30,000 feet.

~ ~ ~ ~

It was a stormy Saturday morning in Sydney (unusual!!) and a rather large and nasty looking CB cloud (cumulonimbus, or thunder cloud) was sitting resolutely over Rozelle, right on the approach path to RW16, the duty runway for that particular day. Several aircraft had reported severe turbulence and shear on final, but in their wisdom, the Tower Controllers had continued with RW16 as the active due to extremely strong crosswinds on RW07. To compensate partially, a severe turbulence advice was being broadcast on the ATIS 115.4 and 132.7. The Sydney Weather Radar was unserviceable and the ILS a bit dodgy due to the heavy rain showers.

F27: "Sydney Tower, we have a CB cell over Rozelle, can we make a brief diversion?"

SYDNEY TWR: "It doesn't look too bad from here and just a few bumps have been reported. Continue approach."

A few seconds pass ...

F27: "Good heavens ... err, Sydney Tower, we just lost control of our aeroplane for a moment there. Please advise following aircraft that there are large rocks in that damn great lump of cloud. Request immediate landing clearance!"

SYDNEY TWR: "Clear to land."

~ ~ ~ ~

A rib-tickling exchange on 133.0, the Apron Clearance Delivery frequency for Perth.

BEECH 200: “Ahh, Perth Clearance Delivery, this is November India Charlie, we’re on the ground Jandakot, request Airways Clearance for Rottnest.”

PERTH ACD: “Roger, November India Charlie, clearance, track Jandakot direct Rottnest, cruise 1500 feet, squawk 4611.”

(30 seconds pass ... and then ...)

BEECH 200: “Ahh, this is November India Charlie, can we have another squawk code assigned to us, something with the second number a zero?”

PERTH ACD: “I guess so, but why, for God’s sake?”

BEECH 200: “Because the knob for the second character has come off in my hand!”

~ ~ ~ ~

A TAA DC-9 was holding prior to descent into Canberra, with the big delay being caused by a flight of 5 Squadron Iroquois helicopters conducting night operations. The helicopter callsigns were ‘Eagle’, with each individual chopper having a different three digit number.

After flying racetracks for thirty five minutes, the skipper of the DC-9 had had enough.

VH-TJJ: “Canberra Approach, this is Tango Juliet Juliet. There’s a little sparrow up there amongst the eagles and he wants to g-e-t d-o-w-n!”

A TAA DC-9

~ ~ ~ ~

Monitored from Qantas 21 late one evening.

“Sydney Sydney, Qantas 21 on five six upper. IFR for Tokyo, preflight check, Selcal Delta Kilo Echo Juliet.”

No answer.

Again, “Sydney, this is Qantas 21, 5643, IFR for Tokyo, Selcal Delta Kilo Echo Juliet.”

Auckland chips in: “Qantas 21, Auckland, go ahead.”

Qantas 21 then recounts the selcal details and gets a successful check.

Auckland then enquires ... “Ah, Qantas two one, Auckland, where exactly are you?”

“Auckland, this is Qantas 21 ... we’re on the ground Sydney!”

AUCKLAND: “Ah ... you wouldn’t have the latest on the football between NSW and Queensland would you?”

QANTAS 21: “Yes ... when we were in Flight Planning it was six-nil NSW’s way!”

AUCKLAND: “Gee thanks Qantas 21, that’s great!”

QANTAS 21: “That’s about the last we heard.”

AUCKLAND: “Alright mate!”

~ ~ ~ ~

The following was heard between an America West 737 (N166AW) and Adelaide control just prior to the pilot's dispute.

N166AW: "Hey Control, Cactus 66 with you ... have you got any weather down there?"

CONTROL: "Roger ... weather outside, but not on radar."

N166AW: "Yair ... well we have a lousy storm up front, can we divert to the left to avoid the activity?"

CONTROL: "Roger, divert as required ... we don't want any white caps in any coffee cups!"

~ ~ ~ ~

New Zealand 115, a Boeing 747-200 completing its last 30 minutes of its Tasman crossing, before turning around the aeroplane with the same crew to return home as New Zealand 116.

ANZ 115: "Qantas Sydney, New Zealand 115, we'll be on the blocks at four seven, we've got a problem with the forward exit door, even though it's closed and locked, it's not recessing on the outside properly. We have no sick people on board, or unaccompanied minors, and we'd like our return passenger load figures for the New Zealand 116 please."

QANTAS SYDNEY: "New Zealand 115, Qantas Sydney, copy your door engineering problem, and your other remarks. Return load is 11 first class, 53 Pacific, and 112 economy."

ANZ 115: "Right! Thank you ... and err ... ah ... our Captain, Captain Freebird ... it's his last flight for Air New Zealand, and he has his wife on the aeroplane and she's returning with us on New Zealand 116. Could you get one of the staff from the boarding lounge to run in on arrival and give Mrs Freebird her return boarding pass, as she doesn't wish to leave the aeroplane, she won't be disembarking."

QANTAS SYDNEY (trying to be helpful): "Er ... New Zealand 115, would you be kind enough to enquire of the Captain would Mrs Freebird prefer a smoking or a non smoking allocation for her return trip."

CAPTAIN FREEBIRD: "Qantas Sydney, this is Captain Freebird. Look ... I don't care what you allocate her, she's going in First Class, non smoking! And that's that!"

~ ~ ~ ~

On a recent flight on a Japanese airline, the Captain illuminated the seat belt warning sign which was followed by an announcement from a cabin attendant. After a long spiel in Japanese which included a warning on expected turbulence, the English translation followed. Short and to the point ... "Ladies and gentleman ... *ah'*... the Captain has illuminated the ... *ah'* ... seat belt warning sign ... please *ah'* ... stand by for terminance!"

A look around the cabin at the other English speaking passengers showed they were all losing a little colour from their faces!

~ ~ ~ ~

Here are some maintenance complaints apparently submitted by US Air Force pilots and the replies from the maintenance crews. "Squawks" are problem listings that pilots generally leave for maintenance crews.

Problem: "Left inside main tire almost needs replacing."
Solution: "Almost replaced left inside main tire."

Problem: "Test flight OK, except autoland very rough."
Solution: "Autoland not installed on this aircraft."

Problem 1: "No 2 Propeller seeping prop fluid."
Solution 1: "No 2 Propeller seepage normal."
Problem 2: "Nos 1, 3 and 4 propellers lack normal seepage."

Problem: "The autopilot doesn't."
Signed off: "IT DOES NOW."

Problem: "Something loose in cockpit."
Solution: "Something tightened in cockpit."

Problem: "Evidence of hydraulic leak on right main landing gear."
Solution: "Evidence removed."

Problem: "DME volume unbelievably loud."
Solution: "Volume set to more believable level."

Problem: "Dead bugs on windshield."
Solution: "Live bugs on order."

Problem: "Autopilot in altitude hold mode produces a 200ft/min descent."
Solution: "Cannot reproduce problem on ground."

Problem: "IFF inoperative."
Solution: "IFF inoperative in OFF mode."

Problem: "Friction locks cause throttle levers to stick."
Solution: "That's what they're there for."

Problem: "Number three engine missing."
Solution: "Engine found on right wing after brief search."

~ ~ ~ ~

Today's flight age is an era highlighted with increasing emphasis on safety. Instrumentation in the cockpit and in the traffic control tower has reached new peaks of electronic perfection to assist the pilot during takeoffs, flight and landings. For whimsical contrast to these and other marvels of scientific flight engineering, it is perhaps opportune to remind pilots of the basic rules concerning the so-called Cat and Duck Method of Flight, just in case something goes wrong with any of these new fangled flying instruments you find in today's aircraft.

Place a live cat on the cockpit floor. Because a cat always remains upright, he or she can be used in lieu of a needle and ball.

Merely watch to see which way the cat leans to determine if a wing is low and, if so, which one.

The duck is used for the instrument approach and landing. Because any sensible duck will refuse to fly under instrument conditions, it is only necessary to hurl your duck out of the plane and follow her to the ground.

There are some limitations to the Cat and Duck Method, but by rigidly adhering to the following check list, a degree of success will be achieved.

- Get a wide awake cat. Most cats do not want to stand up at all, at any time. It may be necessary to get a large fierce dog in the cockpit to keep the cat at attention.
- Make sure your cat is clean. Dirty cats will spend all their time washing. Trying to

follow a cat licking itself usually results in a tight snap roll, followed by an inverted (or flat) spin. You can see this is very unsanitary.

- Old cats are best. Young cats have nine lives, but an old used-up cat with only one life has just as much to lose as you do and will therefore be more dependable.
- Beware of cowardly ducks. If the duck discovers that you are using the cat to stay upright – or straight and level – she will refuse to leave without the cat. Ducks are no better on instruments than you are.
- Be sure the duck has good eyesight. Nearsighted ducks sometimes will go flogging off into the nearest hill. Very shortsighted ducks will not realise they have been thrown out and will descend to the ground in sitting positions. This manoeuvre is quite difficult to follow in an aeroplane.
- Use land loving ducks. It is very discouraging to break out and find yourself on final approach for some farm pond. Also, the farmers there suffer from temporary insanity when chasing crows off their fields and will shoot anything that flies.
- Choose your duck carefully. It is easy to confuse ducks with geese because many water birds look alike. While they are very competent instrument flyers, geese seldom want to go in the same direction you do. If your duck heads off for a swamp, you may be sure you have been given the goose.

~ ~ ~ ~

Oh ... the impact cricket has on us! (Even while flying an F-111 doing touch-and-goes ...)

AMBERLEY TWR: "Macho 891, radar departure, turn right heading 3-1-zero, climb to 2-5-zero-zero, clear for takeoff!"

F-111C: "Macho 891, cleared, climbing to 2500."

APPROACH: "Macho 891, identified ... call final."

APPROACH: "GCA ... Macho 891, 10 miles."

F-111C: "Ahh ... haven't got an update on the cricket have ya?"

APPROACH: "Standby ... one coming through."

APPROACH: "West Indies 3 for 77!"

F-111C: "Roger ... and how many overs?"

APPROACH: "19"

F-111C: "Very, very good!" (said with typical Aussie enthusiasm!)

APPROACH: "Macho 891 ... five miles ... call the Tower!"

~ ~ ~ ~

The following takes place at RAF Gutersloh, Germany in the early sixties when the mini Berlin crisis was on. A squadron of Gloster Javelin fighters was flown in. A fellow was in the crash rescue crew at the time, doing night time security patrols around the flightline.

One night he was out with another firey doing the rounds. The Javelins were armed with the Red Top heat seeking missiles, of which they were quite ignorant, as they had never seen them before. Anyhow, the firies were discussing how these missiles could pick up the heat of a man's body some hundreds of yards away. They had driven the fire truck around the front of the flightline when his mate decided to light up a fag. All of a sudden there was a loud bang and hissing noise.

"Shit," said one, "the bloody missile has picked up the heat of your fag."

With that the driver slammed on the brakes, and they both dived out of the truck, expecting a bloody great missile to be coming straight at them at very high speed!

But after what seemed like ages and nothing had happened, they picked themselves off the ground and cautiously looked around, only to see that they had run over a CO_2 fire extinguisher and had set it off. Luckily no-one had seen them.

Gloster Javelin with Firestreak missiles under the wings.

Bill Lines

~ ~ ~ ~

You all know about the Darwin Awards – it's an annual honour given to the person who did the gene pool the biggest service by killing themselves in the most extraordinarily stupid way or who have performed stupid acts but die without bearing offspring.

The 1995 winner was the fellow who was killed by a Coke machine which toppled over on top of him as he was attempting to tip a free can of drink out of it.

The legendary 1996 winner appeared earlier in this book and was the air force sergeant who attached a JATO unit to his car and crashed into a cliff several hundred feet above the roadbed.

And now, the 1997 winner: Larry Waters of Los Angeles – one of the few Darwin winners to survive his award winning accomplishment. Larry's boyhood dream was to fly. When he graduated from high school, he joined the US Air Force in hopes of becoming a pilot. Unfortunately, poor eyesight disqualified him. When he was finally discharged, he had to satisfy himself with watching jets fly over his backyard.

One day Larry had a bright idea. He decided to fly. He went to the local army disposals store and purchased 45 weather balloons and several tanks of helium. The weather balloons, when fully inflated, would measure more than one metre across.

Back home, Larry securely strapped the balloons to his sturdy lawn chair. He anchored the chair to the bumper of his Jeep and inflated the balloons with the helium. He climbed on for a test while it was still only a few feet above the ground.

Satisfied it would work, Larry packed several sandwiches and a six pack of beer, loaded his pellet gun – figuring he could pop a few balloons when it was time to descend – and went back to the floating lawn chair.

He tied himself in along with his pellet gun and provisions. Larry's plan was to lazily float up to a height of about 30 feet above his backyard after severing the anchor and in a few hours come back down.

Things didn't quite work out that way.

When he cut the cord anchoring the lawn chair to his Jeep, he didn't float lazily up to 30 or so feet. Instead he streaked into the LA sky as if shot from a cannon. He didn't level of at 30 feet, nor did he level off at 100 feet. After climbing and climbing, he levelled off at 11,000 feet. At that height he couldn't risk shooting any of the balloons, lest he unbalance the load and really find himself in trouble. So he stayed there, drifting, cold and frightened, for more than 14 hours.

Then he really got into trouble. He found himself drifting into the primary approach corridor of Los Angeles International Airport (LAX). A United pilot first spotted Larry. He radioed the tower and described passing a guy in a lawn chair with a gun. Radar confirmed the existence of an object floating at 11,000 feet near the airport.

LAX emergency procedures swung into full alert and a helicopter was dispatched to investigate. LAX is right on the ocean. Night was falling and the offshore breeze began to flow. It carried Larry out to sea with the helicopter in hot pursuit. Several miles out, the helicopter caught up with Larry. Once the crew determined that Larry was not dangerous, they attempted to close in for a rescue but the draft from the blades would push Larry away whenever they neared.

Finally, the helicopter ascended to a position several hundred feet above Larry and lowered a rescue line. Larry snagged the line and was hauled back to shore. The difficult manoeuvre was flawlessly executed by the helicopter crew. As soon as Larry was hauled to earth, he was arrested by waiting members of the LAPD for violating LAX airspace. As he was led away in handcuffs, a reporter dispatched to cover the daring rescue asked why he had done it. Larry stopped, turned and replied nonchalantly, "A man can't just sit around".

Let's hear it for Larry Waters, the 1997 Darwin Award Winner.

~ ~ ~ ~

This little gem was written by an American schoolboy to a newspaper a few years ago.

"When I grow up I want to be a pilot because it's a fun job and easy to do. That's why there are so many pilots flying around these days.

"Pilots don't need much school, they just have to learn to read numbers so they can read instruments. I guess they should be able to read road maps too, so they can find their way if they get lost. Pilots should be brave so they don't get scared when it's foggy and they can't see, or if a wing or an engine falls off, they should stay calm so they'll know what to do.

"Pilots have to have good eyes to see through the clouds and they can't be afraid of lightning and thunder because they are so much closer to them than we are.

"The salary pilots make is another thing I like. They make more money than they know what to do with. This is because most people think that flying is dangerous, except pilots because they know how easy it is.

"I hope I don't get airsick because I get carsick and if I get airsick I couldn't be a pilot and if I couldn't be a pilot then I would have to go to work."

~ ~ ~ ~

The holidaying family of four arrived for check-in in Melbourne (MEL) very early in the day and although the father was obviously a seasoned traveller the excitement shown by the rest of the family made it quite clear their holiday trip from Melbourne to Darwin (DRW) was out of the ordinary. The guy at the check-in counter dutifully issued them their boarding passes for their flight to Brisbane (BNE) with the system automatically producing their onward boarding passes for the connecting flight from

Brisbane to Darwin and off they went, no doubt impressed by the efficiency of 1980s airline technology.

It was with some surprise then that later that morning the checkin attendant saw the same family boarding another flight from Melbourne. Apparently, after a delayed arrival in Brisbane, they were paged to quickly proceed to their connecting flight and just made it. It was only after takeoff that the by now slightly frazzled family appreciated that they were heading back to Melbourne. A glitch in the Reservations System had meant they were ticketed MEL/DRW over MEL/BNE/MEL/ADL/DRW sectors.

The father was quite ropeable on his return to Melbourne and the family was upgraded for the remainder of their journey.

~ ~ ~ ~

The following story involves two Ansett A320 aircraft. The situation was like this:

One A320 (we'll refer to it as Ansett 123) was pushing back from an aerobridge at the Perth domestic terminal for an interstate flight, with an arriving A320 (which we'll call Ansett 456) holding on Taxiway Juliet, which is the major entry to the domestic apron for arrivals from Runway 21 and Runway 24. The holding aircraft Ansett 456 was waiting patiently for Ansett 123's bay position.

Normally, an aircraft only takes a minute or so after push back before it taxis, but this day, Ansett 123 sat in the taxi lane for several minutes, with the Surface Movement Controller getting more than a little impatient. He could see a traffic jam developing on Taxiway Juliet, if Ansett 123 didn't move fairly soon from the aerobridge bay.

PERTH GROUND: "Ansett 123, how long until you taxi? Company traffic holding behind you, waiting for your bay."

ANSETT 123: "Uhh ... standby Ground, we have a minor problem."

After another minute or so ....

ANSETT 123: "Ground, Ansett 123, we have to return to the bay. We've left one of the flight attendants behind!"

UNKNOWN: "I guess you would have noticed before now if it was one of the flight deck crew missing?"

Brian Wilkes

Ansett Australia A320

~ ~ ~ ~

The US FAA has a devise for testing the strength of windshields on airplanes called the Chicken Gun. They point this thing at the windshield of the aircraft and shoot a dead chicken at about the speed the aircraft normally flies at it. If the windshield doesn't break, it's likely to survive a real collision with a bird during flight.

The British had recently built a new locomotive that could pull a train faster than any before it. They were not sure that its windshield was strong enough so they borrowed the testing device from the FAA, reset it to approximate the maximum speed of the locomotive, loaded in the dead chicken, and fired. The bird went through the windshield, broke the engineer's chair and made a major dent in the back wall of the engine cab.

They were quite surprised with this result, so they asked the FAA to check the test to see if everything was done correctly. The FAA checked everything and suggested that they might want to repeat the test, this time using a thawed chicken.

~ ~ ~ ~

An Australian couple were in Denver, Colorado in February 1995. This was just before the closure of Stapleton Airport and the opening of its replacement, Denver International Airport, commonly referred to by the locals as DIA.

There had been much public debate and controversy about delays, cost overruns and gremlin removal for DIA's high tech baggage handling system and also mutterings of discontent about the considerably greater distance from the city to the new airport. Thus DIA's opening was about to occur with mixed feelings on the part of the public.

The husband was buying a shirt in a Lakewood (suburb of Denver) department store. Standing next to them at the counter was a couple of senior citizens, presumably locals, obviously husband and wife. The Australian husband expressed some concern to the shop assistant that the shirt may be a little on the large side for him. She said not to worry, just bring it back for exchange if he wasn't happy with the fit. He replied that this may be difficult, as he lived some 20,000 kilometres away. Before the shop assistant could answer, Mr Local turned to his wife and drawled "Probably lives out near DIA." They had a good chuckle at this piece of quick wit!

~ ~ ~ ~

A Bell JetRanger was carrying out some low level circuits in the helicopter training area at the southern edge of Perth's Jandakot airport. Nothing unusual about that, but attached to the underside of the fuselage was some sort of long probe, which extended forward about one metre beyond the rotor-disc.

JANDAKOT TWR: "Helicopter Bravo Echo Lima, what's the probe for?"

BELL 206B: "Ohh ... it's just another 'stinger', we're actually just trying it on a helicopter. Actually, the other day, one of your blokes asked if it was a flagpole!"

JANDAKOT TWR: "Well ... my guess was that it was designed for helicopter jousting!"

Callsigns of aircraft used in this story, apart from Air Traffic Control, were changed to conceal the identities of the aviators.

~ ~ ~ ~

The story involves an Air BC (British Columbia) BAe 146, heard one morning several years ago over western Canada. The flight, which we shall number 'AIR BC 123' was inbound to Vancouver International Airport, cruising at Flight Level 280 and was 75nm (140km) out from its destination.

VANCOUVER CENTRE: "Air BC 123, can you meet a requirement to be at Flight Level 1-8-zero by 60 dme Vancouver?"

AIR BC 123: "Negative!"

VANCOUVER CENTRE: "Don't you guys have a powerful speed brake in the tail that gives you a high rate of descent?"

AIR BC 123: Yes sir, that's affirmative, but it's used for MY mistakes, not yours!"

~ ~ ~ ~

This story involved an aircraft still on the ground at Jandakot, calling on Ground or SMC frequency.

AIRCRAFT: "Jandakot Ground, Delta Uniform Mike, radio check, how do I read?"

JANDAKOT GND: "Delta Uniform Mike, I don't know how you read, but I'm reading you fives!"

~ ~ ~ ~

At one time or another Alice Springs Tower has housed some of the most affable wags ever to keep the planes apart, although sometimes you would have sworn, unless you knew better, that divine intervention was at work. Consider then the sacred and profane aspects of the following events:

It was round about sun up on Sunday, one of those invigorating, gin-clear mornings for which the Centre is noted. Except for a locally based early model Cessna 180, all polished bare metal gleaming in the sun's first rays, its pilot making ready to depart, there was nothing doing within cooee of the place. Said pilot, having settled and rugged up his passengers for a run down to 'The Rock', started up and switched on the ATIS.

Instead of the usual prosaic recorded recital of aerodrome information, what greeted his surprised ears were a few soft bars of church music followed by a familiar voice intoning parsonically, "Here beginneth the lesson. The runway in use is three zero ...." and the rest of it.

Taking his cue on cue, our pilot transmitted thus: "Our father, forget the 'which art' bit, but do give us this day all clearances asked, if that so please you."

Replied the Reverend Controller: "My son, when ready forsake the apron and proceed as you desire, for all the skies are yours to the South and to the West, and none but you, I, and the firery's dog are listening. So may you expect safe passage, and I no retribution."

~ ~ ~ ~

The occasion was an airshow at Amberley in the 1960s, and a small detachment of USAF F-101 Voodoos and F-100D Super Sabres had flown down from Kadena in Japan for the occasion, air refuelled enroute.

The day for the show arrived, and a very young Voodoo jock took to the air to do his thing. Part of his show was a flick roll or two, followed immediately by afterburner light-up and a vertical climb on two columns of hot gas until it was just about out of sight – a very impressive display indeed.

Unfortunately, while over the crowd (black mark!) he lit the burners too soon and the aircraft got very seriously out of shape at very low altitude before he got the nose up. The Voodoo assumed a kind of tumbling attitude from which it smartly recovered.

I was the duty servicing liaison NCO for the visit and met the pilot after he landed. He was as white as a sheet and leant on the wheel and shed a tear or three. When asked just how the hell he recovered from his trouble he replied "Ah just said, it's over to you JC! Ah took mah hands and feet off everything and the bird righted itself!"

Next morning the USAF's detachment's commanding officer, a Colonel, showed me a copy of the Brisbane newspaper, which had a lovely shot of the upside down Voodoo on the front page. He asked me, "Where the hell do I go to buy every goddamned copy of this paper, sarge? Someone is gonna have my ass for this!" They dragged the bird into the hangar, and he did a survey of the airframe so far as was possible to find an unlovely ripple in the 6mm milled alloy skin near the engine bays.

As the Colonel peered at the ripple he asked the seargent, "Waddya reckon sarge? Safe to fly?" he asked.

"If it was mine I'd pull out the bloody engines and instruments and give the rest to the kiddies playground, Colonel!"

Regardless, the Colonel test flew it, going supersonic just to demonstrate to the young jocks exactly how safe it was.

It was surprising to see him back on deck in one piece.

USAF F-101 Voodoo

~ ~ ~ ~

Some years ago a trio of F/A-18 pilots on deployment from Williamtown decided to stretch their wings a bit, and demonstrate a long range supersonic sortie. In the belief that supersonic flight was permitted above FL300, they planned their sortie, blasted out of their deployment base and ran supersonic at FL350 (unconfirmed sources suggest at Mach 1.3) all the way home to Williamtown. Much to their surprise, as the story goes, a reception committee awaited them as they spooled their engines down on the tarmac.

Unfortunately for our intrepid trio of supersonic trailblazers, they failed to consult with their supervising officer, who would have told them that supersonic flight over the Australian continent is not permitted at any altitude. We can probably guess who paid for drinks at the O's Mess that evening!

Editor's Note: Readers may be interested in knowing that both the F/A-18 and the F-111 when clean can cruise at supersonic speeds. Typically afterburner is used to punch through the sound barrier, after which the throttles are backed off either to low afterburning setting or military (maximum dry) thrust at which point the aircraft cruises for as long as the engine temperature rating permits. The F-111 is a particularly slippery performer in this game, and later models of the aircraft with higher thrust engines can sustain a non-afterburning cruise indefinitely, at appropriate altitude, ambient temperature, wing sweep settings and without external stores and pylons. Many stories also abound of overly enthusiastic F-111 pilots burning the paint off the leading edges at Mach 2.6, much to the disdain of the engineering people maintaining the aircraft.

~ ~ ~ ~

This United Airlines story occurred a few years ago, in San Francisco.

Pushy passenger at reservations desk: "Whaddya got goin' to Salt Lake City?"

Reservations agent: "Airplanes and Mormons, Sir!"

~ ~ ~ ~

This was heard recently on a flight into Rockhampton. This was the last flight for the day with the flightcrew staying overnight.

Over the PA system (obviously inadvertently) came: "Bob, that cute blonde flight attendant in the back – tonight I'm going to take her out for dinner, have a shower and then make love to her."

As this was being spoken, a flight attendant was hurrying up the aisle and tripped on hand luggage in the aisle.

As she recovered herself, the elderly lady owning the hand luggage quipped "No need to hurry dear, he said he would have a shower first."

~ ~ ~ ~

Everyone was taking it easy in the HF Room at Gibraltar ATC in the 1950s when they heard this call:

AIRCRAFT: "Gibraltar Tower, Gibraltar Tower, this is Gee-Ay-El-Tee-Y, how do you read over?"

GIB TWR: "Gee-Ay-El-Tee-Y, Gibraltar Tower, reading you 5 by 5."

AIRCRAFT: "Gibraltar Tower, Gee-Ay-El-Tee-Y, roger the 5 by 5, and thank you!"

GIB TWR: "Gee-Ay-El-Tee-Y, Gibraltar Tower, what is your position over?"

AIRCRAFT: "Gibraltar Tower, Gee-Ay-El-Tee-Y is a BOAC Comet Sir, on the ground, Heathrow Airport. Just testing!"

Yes, there is nothing really new in radiospeak, just the alphabet!

~ ~ ~ ~

This story involves an Ansett international flight, Ansett 822, VH-INH, a Boeing 747-300 from Sydney to Kuala Lumpur. The following conversation was monitored, and is one of the "classics" of the airbands!

"Err ... Ansett Operations, Ansett Eight-Two-Two, one of our passengers thinks she has left her iron switched on. Could you please phone her daughter and check if we give you number details?"

An operator at Ansett Operations named John duly phoned the daughter, who apparently said to John in answer to his enquiry of her about the iron .... "Is this some sort of joke?" John reassured her there was no joke, and that it was a genuine message from the captain of Ansett flight 822. "Oh, my mum's a real worry," says the daughter, "hang on, I'll check the iron!"

A moment later the daughter returned. "No, it's okay, tell mum the iron was off". John then asked for the young lady's name so he could verify to the captain that he spoke to the right person. Then John called Ansett 822 back:

"Ansett Eight-two-Two, Ansett Ops, Fiona says the iron is definitely off, and her mother is a big worry!"

When told, no doubt, a relieved mum probably settled back in her seat to enjoy her flight!

~ ~ ~ ~

In the early days of Black Hawk helicopter operations, Air Traffic Control was still coming to grips with helicopters fitted with wheels. A Black Hawk helicopter was coming in for landing at Canberra mid 1988 and the following exchange was overheard between the aircrew and Canberra Tower.

ATC: "Black Hawk A25-102 check wheels."

Black Hawk (after a pause from the pilot and a collective bemused look from the crew):

"Canberra Tower, wheels firmly and securely bolted to airframe."

Another feature of the Black Hawk helicopter that ATC was yet to appreciate was the long legs of the aircraft, when fitted with its four long range ferry tanks. On departure from Canberra heading for Townsville the following exchange occurred.

ATC: "Black Hawk A25-102, request your fuel stops enroute to Townsville."

Black Hawk: "Canberra Tower, none."

Australian Black Hawk

Defence PR

~ ~ ~ ~

Heard on Sydney Tower frequency, and related by an anonymous Navajo Chieftain pilot ....

SYD TWR: "Delta Echo Foxtrot, can you reach five thousand by seven DME?"

VH-DEF: "Shit no!"

~ ~ ~ ~

It was a busy afternoon, and British Airways had been cleared to land on Runway 26. The Tower requested BA to take the highspeed exit off the runway, as there was traffic following.

But on landing, the British Airways aircraft simply stopped on the runway, well short of the highspeed turn-off.

The Tower frantically said "Speedbird, I requested you take the highspeed turnoff!"

A terribly pucka voice replied "Oh, we're turning, we're turning!" He was actually still stationary. The Tower then said "Air Canada One Six, go round Sir, go round!"

As the Speedbird 747 simply ambled off RW26, a very distinct Aussie voice was heard to say: "Up your jumper Speedbird!"

~ ~ ~ ~

This amazing story concerns the old Bristol Freighter – two of the few remaining operators in the '60s and '70s were the RNZAF and the Indian Air Force. The RNZAF Base at Wigram did a fair amount of servicing on these cumbersome old aircraft.

They were puzzled by cracks in the wing roots of the IAF models – unlike any other signs of fatigue experienced in other operators' aircraft. Then all was revealed when an IAF pilot suggested it might have come about during 'looping the loop'! We can't authenticate the story but it makes an amusing tale.

A Bristol Freighter

~ ~ ~ ~

Auckland has an automatic guidance system at the international terminal, and they were having trouble with it on this particular day.

NEW ZEALAND 168: "Auckland Ground, New Zealand 168, pushing back, I don't think we've properly disconnected from the automatic guidance system!"

GROUND: "Yes you have Sir!"

NEW ZEALAND 168: "How do you know?"

GROUND: "We've got you on camera, and can see everything!"

UNIDENTIFIED: "Glad I have my pants on!"

GROUND: "Actually, so are we!"

~ ~ ~ ~

This story concerns a flight attendant for an Australian carrier and one of the first revenue flights of the Boeing 727 in Australia.

The flight began ex Perth and they had a full load. After pre-takeoff checks and the emergency drill etc, the attendants were rushing to their crew positions for takeoff. For some reason our FA's lap strap and shoulder harness were knotted. As she tried very quickly to straighten it out and get belted up the aircraft was caught in a very strong crosswind which caused the air intake on the cabin roof to be briefly denied its essential ingredient, with the result that the engine stalled.

The crew immediately aborted the takeoff, reversing thrust and applying brakes. The sudden change of motion caused the not yet strapped in flight attendant to leave her seat and end up face down in the aisle between two rows of startled and amused passengers! Her pride was the only part that was bruised and she took up her place again to sort out the harness.

The crew taxied to an alternative runway, ran the engines, completed checks etc and proceeded to takeoff, all strapped in and comfortable. All went well, until a latch in the forward galley opened. Sitting inside was a nice silver jug of full cream ready for first class coffee! The jug slid out, and due to the angle of climb, cream poured down the aisle changing the colour of the new carpet on this brand new aircraft!

Once able to commence duty the crew cleaned up and enjoyed an otherwise smooth flight east.

~ ~ ~ ~

One of the pluses of having a parent in the air force when we were young was the fact that on occasions we would get free flights around the country to visit the relatives.

Back in the late 1960s my father was in the RNZAF. My brothers and myself were all under the age of ten and we used to get flights in Bristol Freighters and all sorts of interesting aircraft. A flight from Auckland to Christchurch would take all day as the aircraft would stop at every airport on the way to drop off or pick up pax and freight.

The RNZAF was like a big family then and everyone knew each other and sometimes the flightcrew would have us kids up front. One pilot asked my younger brother to keep an eye open for a petrol station as the aircraft was short on fuel. I remember him standing at the window eyes wide open looking a bit stressed because we were too high to see any stations.

Another time we took off from Dunedin in a Herc, heading back to Auckland the aircraft was diverted to Christchurch. After landing there the crew told us we were home. We realised that we weren't when we walked off the ramp and about six of us tried to tell the crew that this wasn't Auckland. Two six year olds had to sit in the cockpit all the way back to Auckland showing the crew the way home.

I have seen small friends sitting at the controls unable to see over the yoke holding on to the controls with a death grip frozen to the spot 'flying' the aircraft while the pilot has something to eat. One school mate told me that he and several others had to try and push start a Bristol Freighter but it was too heavy. The crew would also often ask one of the kids to count the propellers on one of the engines to see if they were all there. After a few minutes he would return saying they were going too fast to which the pilot would reply that he would turn the engines off to make it easier. "No we'll crash!" was the usual wide eyed response!

~ ~ ~ ~

The three players in this act are Canberra Tower (CB TWR), a Cessna 150 (VH-AWO), and a RAAF Boeing 707, callsign 'Aussie 624'. The prevailing wind was split between Runways 17 and 12, and was blowing at a decent 20 to 30kt (37-56km/h), gusting up to 35kt (65km/h). Clearly, there was a sizeable amount of crosswind on both runways. Our story opens just after the 707 executes a landing on Canberra's Runway 17.

TOWER: "That looked like a Captain's landing!"

AUSSIE 624: "Aussie 624, you bet, I wouldn't trust anyone else!"

A short while later, the Cessna 150 lands on Runway 12.

TOWER: "Alpha Whiskey Escar, no need to acknowledge, vacate the runway as soon as possible, taxiway right."

VH-AWO: "Alpha Whiskey Oscar, wilco, that was a Captain's landing too!"

TOWER: "You guys, you're all showing off today."

VH-AWO: "Yeah, you have to in this wind!"

~ ~ ~ ~

A dull cloudy day at the Brisbane GA airfield of Archerfield, with terminal information 'Echo' reporting three octas of cumulus cloud at 3000ft, and warning of 'Charlie Bravo' (Cumulonimbus, or thunder clouds) approaching from the west. The stars of this story are Archerfield Ground, and VH-JOO, a Cessna 150.

Without further ado ....

VH-JOO: "Archer Ground, good afternoon, Juliet Oscar Oscar for the eastern training area, Runway zero-four-right, and I've received information Echo."

ARCHER GND: "Juliet Oscar Oscar."

This was very quickly followed by ....

VH-JOO: "Archer Ground, Juliet Oscar Oscar, could you give me some sort of estimate as to when to expect the Charlie Bravos to arrive in the vicinity of the airfield?"

ARCHER GND: "Juliet Oscar Oscar, I don't know, could be today, or could be tomorrow!"

VH-JOO: "Well that sure was precise!"

ARCHER GND: (vaguely uncomfortable) "Well ... I mean, they're moving at the moment, they might just skirt the airfield. Actually they might well get here before dark, or maybe after. Oh look ... I don't really know!"

VH-JOO: (said with great sincerity) "With people like you in the Tower, my friend, our airways will always be safe!"

ARCHER GND: "Stop it! I'm blushing!"

~ ~ ~ ~

Here's the Pilot's Prayer, courtesy of "Bill", who is an Air Traffic Controller.

"Oh Controller, who sits in the Centre,
Hallowed be thy Sector,
Thy traffic come, thy instructions be done,
On the ground, as they are in the air,
Give us this day our radar vectors,
And forgive us our CTA incursions,
As we forgive those who cut us off on final,
And lead us not into adverse weather, But deliver us our clearances.
Roger ...."

~ ~ ~ ~

The following conversation was monitored between the RAAF's 'Trojan 204' (a Lockheed C-130 Hercules), Flight Service, and an anonymous "other".

FLIGHT SERVICE: "Trojan two-zero-four, this is Brisbane, request your groundspeed."

TROJAN 204: "Brisbane, Trojan two-zero-four, two-seventy knots."

About a 20 second pause followed ... then ....

FLIGHT SERVICE: "Trojan two-zero-four, Brisbane, ahh ... your time interval to next position is 44 minutes. Can you confirm?"

Thirty seconds passes ....

TROJAN 204: "Brisbane, Trojan two-zero-four, sorry about that, time interval should be 75 minutes, must be a mistake on the flight plan."

FLIGHT SERVICE: "Two-zero-four, many thanks Sir."

About 10 seconds later in a slow country drawl ....

ANONYMOUS: "One plus one equals two!"

FLIGHT SERVICE: "Yeah, good help is hard to find these days!"

~ ~ ~ ~

This story involves Boeing 747 freighter N470EV from Evergreen wet leased to Qantas for freight. On this occasion, it was operating as 'Evergreen 103' at Melbourne International Airport. The freighter had just been given clearance to line up and hold on Runway 34, when the following conversation broke loose:

EVERGREEN 103: "Tower, Evergreen one-zero-three, ready, runway three-four."

MELBOURNE TWR: "Evergreen one-zero-three, roger, hold position, I've got one landing on Runway two-seven before you roll."

EVERGREEN: "Evergreen one-zero-three, holding position."

Shortly after ....

MELBOURNE TWR: "Evergreen one-zero-three, I've got a message from your company. They want you to call in before you go. Apparently there's a problem with the loading. It might be better to just wait there and talk to them."

EVERGREEN 103: "Okay Sir, we'll call them right now, Evergreen one-zero-three."

The following conversation then erupts on Qantas Operations frequency:

EVERGREEN 103 "Qantas Ops, Evergreen one-zero-three."

QANTAS OPS: "Evergreen one-zero-three, this is Qantas Melbourne. Thanks for calling Sir. Err ... your aircraft has not been loaded up! There are apparently two more pallet stacks that have to be loaded on your aircraft."

EVERGREEN 103: "Are you requesting that we return to the gate?"

QANTAS OPS: "Affirmative Sir, if you taxi back to the gate, we can fully load your flight."

EVERGREEN 103: "Okay Sir, we're coming back."

QANTAS OPS: "Evergreen one-zero-three, Qantas Melbourne, copy Sir, and thank you."

Now back on Tower frequency 120.5 MHz:

EVERGREEN 103: "Ahh ... Tower, this is the Evergreen one-zero-three, we're going to have to return to the gate Sir!"

MELBOURNE TWR: "Evergreen one-zero-three, roger. I'd like you to taxi forward along the runway and vacate the runway at Taxiway Juliet."

EVERGREEN 103: "Forward, and vacate at Juliet, Evergreen one-zero-three."

Vacating the runway ... Evergreen 103 changed to Melbourne Ground frequency on 121.7 MHz.

EVERGREEN 103: "Melbourne Ground, Evergreen one-zero-three, we're now clear of the runway via Juliet, and I guess they want us to go back to the cargo ramp?"

MELBOURNE GND: "Evergreen one-zero-three, yes, back into Golf-2 cargo ramp position. Taxi straight ahead via Juliet, onto the apron, and then into Golf-2."

EVERGREEN 103: "Okay Ground, Juliet through to the apron, and then Golf-2."

The huge Evergreen 747 freighter returned to its allocated parking position, and the remaining two cargo pallets were hastily loaded, resulting in a one hour delay.

~ ~ ~ ~

Another Evergreen story worth telling, this time exclusively involving Melbourne Ground and Evergreen.

EVERGREEN 103: "Ground, Evergreen one-zero-three, all ready for taxi."

MELBOURNE GND: "Evergreen one-zero-three is clear to taxi Runway two-seven, wind now light and variable, occasional downwind of five knots, QNH one-zero-one-three, temperature one-four, and the time is four-three-and-a-half."

EVERGREEN 103: "Okay, we're headed for Runway one-seven, and copy all the weather, Evergreen one-zero-three."

MELBOURNE GND: "Evergreen, that was Runway TWO-SEVEN if that's acceptable, and take first taxiway left onto Yankee, and then hang a right onto Sierra."

EVERGREEN 103: "Left on Yankee, then right on Sierra, and that's to Runway one-seven. My fault Sir, sorry" (thinking he had corrected his previous mistake).

MELBOURNE GND: "No ... no ... no! Runway TWO-SEVEN."

EVERGREEN 103: (Laughing) "Ahh, sorry Sir, that should be Runway two-seven! I just can't seem to get it through my head!"

An Evergreen 747-100 Freighter

Craig Fraser

~ ~ ~ ~

Occasionally Brisbane has cold, wet and windy winter nights and on this particular evening (quite some time ago) there was a 707 overnighting that had a stabiliser trim motor problem. The stab trim motor is located behind the rear pressure bulkhead in the unpressurised area under the fin, (ie: way back and way up).

Two engineers were in that area, accessed through a small panel and by a tall stand fitted with wheels to make handling easy, while another was on the flightdeck operating controls and reading maintenance manual procedures to the tail end men via the intercom.

It was 0400hrs and the security guard decided to stay inside the warm terminal building rather than out on the icy ramp and had managed to fall asleep.

The stab trim motor was not behaving itself as the manual said it should so the engineer left the flightdeck to join the tail end guys. Climbing the stand he squeezed through the access panel for discussions and adjustments. Satisfied they were now ready for functional checks the engineer lowered himself feet first through the entry hatch as far as his armpits but searching with his feet could not find the platform. Climbing back in and looking out revealed the wind had blown the stand over 15 metres away. Urgent shouting and waving torches failed to attract anyone's attention, especially the sleeping guard's.

Desperate situations require desperate solutions! They all removed their jackets and overalls and tied them into an escape rope. Two of them held one end while the nominated volunteer was lowered towards the ground. On reaching the full length of the chain he then had to drop the final metre and bring the stand back for the rest of them to escape!

~ ~ ~ ~

In 1988, an Australian was serving with the US Air Force in Virginia, and took his small six year old son to the local Family Open Day at Langley Air Force Base (AFB) which is home to a squadron of F-15 Eagles.

Security was relaxed, due to the family nature of the day, and displayed aircraft were not roped off, and were fully accessible to the crowd. It was a hot steamy day, with families sheltering under wings, sticking their heads into exhaust pipes etc.

Our young father sat his son in the shade on the bottom lip of an F-15 port engine intake. It was a factory-fresh F-15, with stairs up to the cockpit, and a young enlisted airman at the top, supervising visitors, who were able to visit the cockpit.

The father and son were thrilled at the chance of being able to sit in a flashy fighter, and lined up in the hot sun for their turn. After a 15 minute wait, they reached the ladder top, and the father gently lowered his six year old son into the pilot's seat, with the young supervising airman close by.

At this point, destiny took over!

The young fella had only just learnt to read, and could only handle single syllable words at that stage, but that was enough. The little boy looked down to his right and read the word "PULL" on a neat little T-handle. And pull he did!

Time stood still. Deep in the bowels of the big fighter jet, a rising whine was heard, and the young American supervising airman went white!

Highly nervous fathers retrieved offspring from engine intakes, while others removed their heads from afterburners. Fortunately the whine died, presumably a servo-motor operating on battery power, that shut down for want of further current. But it was enough to issue a fright and cause panic.

The embarrassed father and six year old mischief maker left. In a rather big hurry.

~ ~ ~ ~

An oldie but a goodie:
MIAMI CENTRE: "November-1-2-3-Yankee-Zulu, say altitude."
N123YZ: "Altitude!"
MIAMI CENTRE: "November-1-2-3-Yankee-Zulu, say airspeed."
N123YZ: "Airspeed!"
MIAMI CENTRE: "November-1-2-3-Yankee-Zulu, say Cancel Clearance!"
N123YZ: "Eight thousand feet Sir, one hundred and fifty knots IAS!"

~ ~ ~ ~

In the 747, as with all modern airliners, toilet tanks are inbuilt into the aircraft so waste can be dumped into specially designed transporters while on the ground (it does not go overboard while in the air).

To allow dumping a special valve is fitted in the tank which is actuated by a mechanical lever externally located on the fuselage underside. Each toilet has its own tank except the two rear toilets which share a common tank.

One particular aircraft was new in service and a new engineer, who had no previous experience in this facet of servicing, was taken by the training engineer to the rear toilet where he was fitted with the regulation length armpit glove, instructed to kneel in front of the bowl and insert his arm down into the tank to feel for the dump valve and locate the obstruction that was causing a dump problem.

Meanwhile another engineer entered the other rear toilet (fitted with a glove) and putting his arm down the bowl grabbed a hold of the trainee's hand.

Frequent travellers will know that the toilet door opens inwards and closes by outward pressure.

Having now encountered the unknown monster from the depths of the tank the trainee was on his feet and frantically exiting through the closed door before one could blink an eye!

Unfortunately the aircraft was delayed while repairs were made to the toilet's door which had "been found damaged" during the transit.

~ ~ ~ ~

While an Aussie was based on an overseas station as a LAME, an incoming aircraft radioed ahead during a night transit that a passenger had lost a ring in a toilet and could they locate it please.

As these were pre environmentally conscious days the toilet operator simply dumped the waste tank onto the tarmac. A search by torchlight while walking around the perimeter failed to find the errant ring.

Just as they had decided to wind up the search and clean up the mess a male passenger dressed in a pale blue suit and white shoes arrived and immediately started crawling on his hands and knees in desperate search of the ring.

At this point they realised they were not looking for "just a ring".

His search proved fruitless and as he was approaching panic point, the operator was instructed to go upstairs and search the tank.

A few minutes later he arrived with a block of gold about four centimetres cube with a hole through the middle big enough for a finger to fit through.

"Thanks" said the passenger, as he pushed the unwashed ring on his unwashed finger and headed off in the direction of the terminal with a grin from ear to ear.

~ ~ ~ ~

This tale occurred during the 1970s on a Fokker Friendship F27 between Melbourne and Mt Gambier on a night service. The flight had been enjoyable, smooth and incident free until they prepared to land in Mt Gambier at about 9.00pm.

The aircraft began to descend further towards a long line of white lights, when suddenly the engines were roaring to full power, the undercarriage was raised, and they gained altitude again.

The aircraft circled around and again approached a long stream of lights, different ones! Power was reduced, the gear put down and they made a smooth landing. The passengers thought the aircraft had to circle due to traffic congestion, which was partly true, as a rather embarrassed crew explained afterwards that they had lined up on the main road out of Mt Gambier on their first approach, not the airport!

~ ~ ~ ~

It was a cold and bleak day early in WW2. A Royal Navy squadron operating Avenger ASW aircraft, which had been based by itself at a lonely naval airbase in the north west of Scotland, had suddenly been joined overnight by a squadron of Martlets. Now as any aviation Buff (Big Ugly Fat Fellow) worth his salt knows, the Avenger and the Martlet were designed and built by perhaps the greatest manufacturer of naval aircraft, Grumman. As a result, there was strong family resemblance between the two aircraft, although the Martlet, being a fighter, was much smaller in size.

On the day in question, an old Chief Petty Officer who had learnt his trade on Fairey Flycatchers, and the much more modern (to him) Swordfish, came round the corner of the hangar together with the squadron air engineer officer to behold the little Martlets lined up in front of the Avengers.

Taken completely by surprise, the old Chief turned to the AEO and in what would have to be a classic one liner, exclaimed, "My God Sir, they've pupped!"

The Grumman Martlet was a version of the Wildcat (pictured) used by the Royal Navy during WW2.

~ ~ ~ ~

Overheard on Coolangatta Tower frequency.

COOLY TWR: "Alpha Bravo Charlie, report position."

VH-ABC: "Alpha Bravo Charlie, in your two o'clock."

COOLY TWR: "Alpha Bravo Charlie, err, roger, Cooly Tower, but what way am I facing?"

~ ~ ~ ~

Essendon Tower, early one morning. The Police helicopter, POLAIR 1 was airborne, near to the Melbourne downtown area, and had requested an inbound clearance to Essendon.

The wind conditions were blustery, making light aircraft and helicopter flying 'uncomfortable' to say the least.

ESSENDON TWR: "Polair 1, standby, I'd like to descend a 737 over the city for Tullamarine."

POLAIR 1: (Sounding fed-up) "No worries, ask him if he wants to swap jobs!"

~ ~ ~ ~

In 1944 a pilot flying Horsa gliders with the Airborne Forces in Europe was sent back to a regular piston engine refresher course at Larkhill with some other members of his brigade where they flew Tiger Moths.

It was one of those very hazy days where there was no horizon, just the same silvery blue-grey glare above and below. The white ball of the sun, suspended low ahead, was the only clear visual reference. This was par for the course for anyone who remembers flying in that era of pea soup fogs and startlingly clear inversion layers marking the upper limits of the murk. The pilot recalled:

"I was sitting there heading for base, cockpit sides down, goggles up, elbows outside, sitting back enjoying the serenity of it all when suddenly I looked down to starboard and saw my mate a hundred yards or so away doing the same thing in his Tiger. I saw him look up and wave. I waved back. Then I watched enthralled as he slowly came up ... and up ... until he rolled right over me and came down the other side. He stayed there, looked up and waved. I waved back and followed him back to base. He landed just in front of me where I was switching off. He pulled off his helmet slapped me on the shoulder and said to me, "Hey, Hank that manoeuvre was fantastic! How did you do it?'"

~ ~ ~ ~

The dateline is Melbourne International, and the main 'star' is a freight jockey, who was about to taxi from the Australian air Express freight apron, which is right near the Qantas terminal.

VH-ABC: "Melbourne Ground, Alpha Bravo Charlie at Aussie air Express, request taxi clearance, received information Mike."

MEL GND: "Alpha Bravo Charlie, clear to taxi, caution, Qantas 737 coming out from behind the finger."

VH-ABC: (who had just received a knock back from Qantas) "Yeah, that's a good metaphor!"

~ ~ ~ ~

In September 1989, when Australia had the domestic pilot's dispute, an America West Airlines 737, callsign "Cactus six seven", was operating a service for Ansett, and was in the line up position at Melbourne on Runway 27.

MEL TWR: "Cactus six seven, cancel S.I.D, maintain runway heading, climb to three thousand, remain Tower frequency airborne, clear for takeoff."

The reply came in a thick Texan drawl ....

N167AW: "Cactus six seven, we're gone, up to three thousand, straight ahead and stayin' with y'all."

~ ~ ~ ~

This one's from Sydney. An aircraft was waiting to depart Sydney at night (curfew departures) when a National Jet aircraft was inbound.

SYDNEY TWR: "ABC, clear to line up behind the landing 737."

VH-ABC: "Roger."

A minute or so later, a BAe 146 rolls on past ....

VH-ABC: "Ah, Tower, confirm that 737 is actually a 146?"

SYDNEY TWR: "Yes, you're quite right!"

VH-ABC: "No problems, they're easily confused. The only difference is that the 737 has one APU, and the BAe 146 has five!"

~ ~ ~ ~

The form used for Britain's Royal Navy and Marines officer competency reports is the S206. The following are apparently true excerpts taken from people's '206s' ....

- His men would follow him anywhere, but only out of curiosity.
- This officer is really not so much of a has been, but more of a definitely won't be.
- When she opens her mouth, it seems that this is only to change whichever foot was previously in there.
- He has carried out each and every one of his duties to his entire satisfaction.
- He would be out of his depth in a car parked in a puddle.
- Technically sound, but socially impossible.
- This officer reminds me very much of a gyroscope – always spinning around at a frantic pace, but not really going anywhere.
- This young lady has delusions of adequacy.
- When he joined my ship, this officer was something of a granny; since then he has aged considerably.
- Since my last report he has reached rock bottom, and has started to dig.
- She sets low personal standards and then consistently fails to achieve them.
- He has the wisdom of youth, and the energy of old age.
- This officer should go far – and the sooner he starts, the better.
- In my opinion this pilot should not be authorised to fly below 250 feet.
- This man is depriving a village somewhere of an idiot.
- The only ship I would recommend this man for is citizenship.
- Works well when under constant supervision and cornered like a rat in a trap.

~ ~ ~ ~

One from the good old days at Coolangatta.

A Cessna 210 was taxying out for some circuits, but the Tower spotted that it still had its gear doors hanging down. The 210 replied that it'd just had some maintenance done, and the doors would disappear when the gear retracted for the first time. Another pilot who was on frequency at the time thought he'd help out by pointing out on the radio that the 210 pilot could finish the extension cycle, and close the doors by giving the emergency extension handle a few pumps. Cooly Tower came back with this ....

COOLY TWR: "That sounds like Bill Yeagher."

UNKNOWN: "Nah, more like Chuck Up!"

~ ~ ~ ~

Inbound to Wagga, while just skimming under the clouds ....
WAGGA TWR: "Delta Echo Foxtrot, report inflight conditions."
VH-DEF: "Delta Echo Foxtrot, excuse the expression, but I'm in and out of bottoms!"

~ ~ ~ ~

Back in 1976, on a Qantas flight from England on the Bombay/Perth leg, the crew decided to liven up the life of a lonely young man.

Some 30nm (55km) from Cocos Island, the skipper was on radio giving a position report, and when he had finished the report, chatted to the Cocos operator about a number of topics – life on Cocos, how long the operator was stationed there, what was life on Cocos Island like, were there any women?

The operator replied that he didn't have long to go, and that one of the worst aspects of the job was that there were no women. The Captain commiserated with the poor guy, then signed off, saying he would call 'over the top' – he would report his position directly overhead Cocos Island.

The Captain then hurriedly arranged for a senior female Flight Attendant with an especially sultry voice to be summoned to the flightdeck. He then wrote down word for word the 'over the top' Cocos Island position report, and then when approaching overhead, handed the sultry lady the mike and had her deliver the report to our female-companion-starved Flight Service operator below.

Her normally sultry voice became extra low and sexy, and she really got into the part. What followed from the FSO was a series of loud groans, while the whole flightdeck crew rolled with laughter, with tears in their eyes.

~ ~ ~ ~

Our hero in this tale is A copilot on a RAAF C-130E Hercules.

The mission was a standard five day RAAF Butterworth resupply, which went Day 1 Richmond/Darwin; Day 2 Darwin/Butterworth; Day 3 shopping in Penang; Day 4 Butterworth/Darwin; Day 5 Darwin/Richmond.

We pick up the story on Day 5, approaching top of climb out of Darwin for Richmond with the Captain flying:

As a 'boggy' Pilot Officer the crew have had great fun pulling the copilot's leg for four days, cashing in on his inexperience, all of which is taken in good jest. Typically the SNCOs (the Flight Engineer and Loadmaster, known as Scruff and Shirley respectively), are the main antagonists.

Consequently, when Scruff taps the copilot on his shoulder and says "Will you have a look at that bloody big spider", the copilot assumes it's another prank. To our hero's horror, as his gaze is averted from looking out the window and follows the pointed finger, he sees the world's biggest, most hideous looking eight legged creature emerging from a gap in the overhead panel about 20cm from the Captain's face.

Well, the fearless leader must have set a new world record for unstrapping and exiting his seat, leaving the copilot to fly the aeroplane and allowing a close inspection of this 30cm wide, orange coloured monster as it heads his way.

The copilot is by now very concerned about his immediate wellbeing but not in a position to do much about it. 'Fang' as he will be called, is slowly crawling over the fire handles and extinguisher switches for the engines, towards the right hand seat. The crew cannot hit Fang for fear of damaging the systems and are reluctant to use a spray in case it shorts out anything electrical.

Fang looks foreign and as Scruff backs his seat away he comments that the cargo load in Butterworth had been on the apron for some time and he must have come aboard on it. Shirley now says “hey that looks like one of those Malaysian cigarette spiders”.

“What the hell is that?” the copilot asks.

“Well if they bite you *that's* about how long you've got.”

Our hero looks at him and Shirley isn't smiling.

After another two minutes of eyes to eye contact, Fang disappears up another gap above the copilot, this time never to be seen again.

All the crew now think it's a big joke, with the copilot staring at Fang's departure place for the next five hours. Scruff even adds a nice touch by gently brushing his headset cord across his neck. The copilot nearly hits the roof, with the boys rolling around with laughter.

The Captain eventually summoned up enough courage approaching Richmond to carry out the landing. Quarantine at Richmond however, took the issue very seriously and after the crews disembarking, sprayed around enough insecticide to kill a mother Alien.

Finally, the groundies had a look but Fang is never found, his corpse forever lurking in the bowels of C-130E Hercules A97-181's electrical system.

Home to Fang? RAAF C-130E Hercules A97-181

~ ~ ~ ~

Two aircraft were approaching the holding point at Canberra's Runway 35. The first aircraft was an Eastern Dash 8, Sydney bound, right behind it a Kendell Saab, also bound for the same destination.

EASTERN DASH 8 (male): “Canberra Tower, Delta Echo Foxtrot ready.”

KENDELL SAAB (female): “Canberra Tower, Golf Hotel India ready.”

EASTERN DASH 8: “Is that Vanessa?”

KENDELL SAAB: “Yo baby Yo!”

A few moments silence ... then ....

KENDELL SAAB: “Race ya!”

~ ~ ~ ~

A LAME in Cairns was once asked if the tropical climate caused any peculiar aircraft maintenance problems.

He thought for a moment and said, "Yes, cockroaches, because they seem to thrive in all the hidden recesses despite blazing heat on the ground and chilly temperatures in flight, and can really make a mess."

Then he told a story about a businessman who woke up some years ago on an international flight only to find a giant cockroach crawling down his cheek. He was so revolted by this that he filled in a complaint form about the incident and sent if off.

Some weeks later a letter from the airline's Public Relations Manager landed on his desk. It was an outstanding example of the practitioner's art and convincingly explained the airline's strict precautions against such pests before admitting that, in very rare circumstances, bugs did get on board their airliners, but when they did so, it was always somebody else's fault; it was the unhygienic dumps they had to land at, or when they had to rely on contractors for servicing, or when carried aboard in passengers' hand luggage.

Then, to show how repentant the airline was, the PR manager invited the business man to present his letter at check-in when next he flew, when he would be automatically upgraded to first class.

After carefully filing the letter, the business man crumpled up the envelope it came in and pitched it into the waste paper basket. As it landed he noticed something inside it. So he retrieved the envelope and withdrew one of the ubiquitous Post-it notes used nowadays to write messages in every office around the globe, and apparently enclosed in error by the PR manager's secretary.

It was his instruction to her to, "Just send this jerk the standard cockroach letter".

~ ~ ~ ~

One of Australia's most experienced DC-3 pilots is Terry Burns of Hobart.

In 1950 Terry was a first officer with ANA. Taxying out of Sydney one morning for Melbourne Terry thought he saw the door warning light flicker. He was just pointing this out to the skipper when the hostie, known as 'Goldie', put her head into the cockpit and said they'd had a Brisbane passenger on board.

"Had?" said the skipper incredulously.

"Yes", said Goldie, "but I let him out and chucked his bag after him. It was the last one loaded in the rear hold".

"We're done son", said the skipper to Terry.

They turned around at the runway end and in the distance saw a lone figure limping off towards the terminal clutching his bag. At least he hadn't been struck by the tailplane it seemed.

On arrival in Melbourne the two pilots were summoned upstairs for a grilling. The skipper successfully stuck to his argument that the pilots could hardly be expected to control the aeroplane and monitor unsanctioned use of the cabin door.

According to Terry, Goldie survived to fly another day but was eventually dismissed for being late for work one day too many. She had sent five pounds off in a round robin and would wait full of hope each morning for the postie to arrive with the windfall that never came.

~ ~ ~ ~

Unfortunately, some night freight aircraft being flown don't have autopilots, with the result the aeroplane tends to 'wander around' a bit, some of the time.
BRIS CENTRE: "Alpha Bravo Charlie, confirm tracking direct to Jacob's Well?"
VH-ABC: "Affirm, and I'm also trying to sign my name on your radar screen. How am I doing?"

~ ~ ~ ~

Two local charter aircraft were returning to Darwin late on a Friday afternoon, and they obviously knew each other, and the air traffic controller.
TOWER (133.1): "Alpha Bravo Charlie, you're number one, expect left base, Runway one-one."
Then a short time later ....
TOWER: "X-ray Yankee Zulu, you're number two, follow company aircraft Alpha Bravo Charlie in your 10 o'clock position report sighting."
VH-XYZ: "Traffic sighted, we'd like to pass him and track in as number one."
TOWER: "Alpha Bravo Charlie, company aircraft is X-ray Yankee Zulu, in your four o'clock position, report sighting."
VH-ABC: "Traffic sighted."
TOWER: "X-ray Yankee Zulu wants to pass you Alpha Bravo Charlie, so is it okay for you to track as number two."
VH-ABC: "Okay, as long as he buys me a beer!"
TOWER: "Alpha Bravo Charlie, follow company aircraft X-ray Yankee Zulu. Break – X-ray Yankee Zulu, track as number one, left base for Runway one-one, ATC requirement, that you buy Alpha Bravo Charlie a beer on the ground Sir, and also one for me for that matter!"

~ ~ ~ ~

You may not realise it, but the reason Ansett selected a series of 'Eights' in its callsigns for its flights to Japan, Hong Kong and other Asian services is very simple:
Eight is a lucky number to many Asian people.
So the use of as many eights as possible in the callsign numerals is a great marketing spin for Asian passengers booking on Ansett flights. But all the eights creates something of a mouthful for aviators, and on a busy Friday night in Hong Kong recently, the Tower asked an Ansett flight whether they were ready for an immediate departure.
The intended response was "Ansett Eight Eight Eight, affirm".
It came out as "Ansett Eight Eight Eight A......Arghh, oh gee, err, Ansett is ready immediate!" They were a bit behind the eight ball.

~ ~ ~ ~

Canberra has a number of hills in the surrounding terrain which sometimes contributes to strange radar returns.
In an aircraft about six miles out of Canberra at 4000ft, the following occurred:
APPROACH: "Alpha Bravo Charlie, Canberra Approach, there is traffic at 12 o'clock, two miles, altitude unknown. I think it might be a truck on the highway!"
VH-ABC: "Okay, we'll keep a lookout for anything that moves!"

~ ~ ~ ~

Living in England during the 1970s and early 1980s provided opportunities to attend a wide range of open days and airshows at RAF and USAF bases. One of these was the 1979 International Air Tattoo at RAF Greenham Common (before it became infamous as a USAF cruise missile staging post), where I witnessed an amusing illustration of how armed forces can be a very accurate mirror of the national character of the country they defend.

One of the features of IAT '79 was a fly-in from air forces operating the C-130 Hercules, to mark the 25th anniversary of the Lockheed airlifter's first flight. The sight of tens of Hercules lined up, wing tip to wing tip and in alphabetical order of their country of origin, beside Greenham Common's 3.5km long main runway, was impressive to say the least.

One of the first Hercs to have a crew in attendance was the USAF's. Backed by a series of boards illustrating the range of missions the Hercules could fill, the crew, dressed in combat fatigues with name patches and almost invisible badges of rank, proudly and loquaciously described how the Hercules and other airlifters in the USAF inventory gave Uncle Sam the capability to rapidly deploy military forces to anywhere in the world, in defence of peace and freedom loving peoples everywhere.

Nearby was the UK, with the RAF chaps in flying gear and pullovers with prominent rank boards enthusiastically describing what a wizard kite the Hercules was to every passing schoolboy. Further up were the Israelis, wearing sidearms and formed in a semicircle facing out from the aircraft, midway between it and the crowd line. They were constantly scanning the crowd, talking to nobody, suspicious of everyone.

Then I came to the Brazilians, all dressed in immaculately pressed camouflage fatigues, with bright yellow neckerchiefs at the collar, berets all set at an identically rakish angle, gazing disdainfully at the passing parade through dark Ray-Ban aviators – unless, of course, you were blonde, female and reasonably attractive.

Finally I approached the head of the line, and experienced a glow of national pride as I saw the first Herc bore a small dayglo orange kangaroo beside the front hatch. I was a little disappointed at first to see that the crew weren't in attendance. Then I looked up and spotted our brave aviators – all topless on top of the wing, lying back, soaking up the English summer sun!

~ ~ ~ ~

It was Christmas dinner at Coomalie airfield in the Northern Territory in the airmen's mess during WW2 and the orderly officer came around to ask were there any complaints. An airman stood up and said "Sir, the chicken is so tough I think it has had more flying hours than a Wirraway".

~ ~ ~ ~

"Stewardess!"

With the sun beginning to rise, the cabin of the 747 was suddenly illuminated.

"Who turned on the f...... lights?" a male passenger, who had been surly since boarding, snarled at a flight attendant.

The girl had had enough of this particular character.

"These are the breakfast lights, Sir," she answered with forced sweetness. "The f...... lights are much dimmer, and you Sir, snored right through them".

~ ~ ~ ~

The following radio transmission was heard while flying between Brewarrina and Bathurst in NSW.

VH-AMM: "Sydney Control, Alpha Mike Mike, inbound to Bathurst at Flight Level 1-9-zero, overhead Dubbo. My ETA is four-four."

CONTROL: "Alpha Mike Mike, Sydney Control, IFR traffic in your area is a Shorts 360, Xray Yankee Zulu, doing circuit training at Bathurst, and another Shorts 360, Alpha Bravo Charlie, inbound from Orange. ETA is three-six."

UNKNOWN: "I guess you'll have to keep your eyes open for a pair of shorts!"

~ ~ ~ ~

At Sydney Airport one day there was a Cessna 172 waiting to takeoff with a Qantas 747-400 behind it.

Slowly the Cessna rolled down 16R and took off climbing slowly. The Qantas Captain was obviously eager to depart so he asked Control if he could roll yet. Control's response came back "Negative". The Qantas Captain was now a bit confused, because the Cessna was way clear of any aircraft. He asked Control "Why can't we depart?" Control retorted "Hold – wake turbulence Cessna 172!"

Cessna 172

~ ~ ~ ~

Here is one snippet from a Flight Advisory Notice from HQADFOps, relating to US President Bill Clinton's recent visit to Australia.

Message content:

"The following flight details are forwarded for your consideration. Orig: Hickam – Destination: Richmond. A. Operator: USAF. B. Type: C-5. C. Tail: 870036. D. Callsign: REACH 72589. E. Captain: TBA (To be announced). F. Crew: 9. G. Pax: TBA. H. Purpose: Presidential visit support-reserve, backup, standby, a just in case, you never know when it might come in handy aircraft."

~ ~ ~ ~

This story concerns a trainee Air Traffic Controller at Washington's National Airport USA, with a really stern instructor looking over his shoulder.

A Boeing 737 was just touching down, and he had an MD-80 on a one mile final, with an ATR 72 holding short of the runway for departure.

The trainee decided he could slot the ATR between the jets, to which the instructor replied it would work, but only if nobody fouled up, and if it was done quickly. The training officer finished his assessment of the situation with: "If this works, it will be nothing short of a miracle!"

The trainee picked up the mike and intended to say:

"Commuter 1-2-3, position and hold, be ready for an immediate."

But, what came out was ....

"Commuter 1-2-3, position and hold, be ready for a miracle!"

After what apparently was an interminable pause, the commuter pilot replied:

"Tower, under the circumstances we'd better just hold short. I'm a non believer!"

~ ~ ~ ~

Another good one involves Chicago Meigs Airport Tower. (Meigs is a small, GA, city based airport in downtown Chicago).

TOWER: "November 3-4-Bravo is cleared for touch-and-go, and by the way, that's the 29th!"

Several more circuits later ....

N1234B: "Tower, November 3-4-Bravo, final, for touch-and-go!"

TOWER: "November 3-4-Bravo, you're cleared for Touch-and-Go. Listen brother, how many more circuits were you planning on making?"

N1234B: "3-4-Bravo, just a couple more!"

TOWER: "Roger, err.....ah, I just wondered buddy 'cause we were calculating your landing fees, and you're now up to thirteen thousand, seven hundred and thirty six dollars and eighty-nine cents!"

There was a long delay.

N1234B: "Ahh ... Tower, this is November 3-4-Bravo, that was our last one, we're coming in for a full stop!"

TOWER: "Just kidding brother! Next time, read your Flight Supplement!"

~ ~ ~ ~

In the USA an instructor and his student are holding on the runway for departing cross traffic when suddenly a deer runs out of the nearby woods, and stops smack bang in the middle of the runway, and looks over at them.

TOWER: "Cessna November 6-7-Yankee, clear for takeoff."

STUDENT: "What should I do?"

INSTRUCTOR: "Maybe if you taxi towards the deer, it'll scare him away!"

This is what happens, but the deer is macho, and holds his position, unflinchingly.

TOWER: "6-7-Yankee, clear for takeoff!"

STUDENT: "Tower, November 6-7-Yankee, Sir ... err, there's a deer on the runway!"

TOWER: "Roger sir, hold your position. Deer on runway zero-five, you're cleared for IMMEDIATE departure!"

Unbelievably, two seconds later, the deer bolts from the runway, and runs back into the woods.

TOWER: "November 6-7-Yankee, clear for takeoff, caution wake turbulence, departing deer runway zero-five!"

~ ~ ~ ~

During the early seventies a reservation officer with Ansett answered a phone one morning and the conversation went something like this:

CALLER: "I would like to make a reservation to Jeopardy please."

ANSETT: "Have you any idea where it is as I have never heard of it?"

CALLER: "Neither have I but I have just heard on the radio that there are 300 jobs in Jeopardy and thought I had better try for one."

At this stage it was explained just what was meant about 'jobs in jeopardy'. The reservation agent felt very sorry for the youngster but had to give him full marks for trying.

~ ~ ~ ~

While visiting the United States not too long ago, our correspondent was dutifully monitoring the Tower at Chicago's O'Hare Airport, after a DC-10 skipper from American Airlines made a long landing after coming in too fast. This was the comment from the Tower Controller:

TOWER: American 751 heavy, turn right at the end, if you're able to stop before then. Otherwise take the Midway exit off Highway 101 back to the airport!

~ ~ ~ ~

One dark night in 1945 flying in a battered old Halifax out of Yorkshire with 420 Squadron one particular aircraft was given the extra job of dropping Window, which was the metallic coated strips of paper that interfered with enemy radar (chaff).

A chute extension had to be plugged into an aperture in the floor to enable the Flight Engineer to deploy the bundles from his station. Having done this he waited until the target release time and commenced dropping the window at the recommended rate of two bundles a minute.

Approaching Hamburg, things were hotting up outside, and he logically reasoned that if two bundles a minute kept them reasonably safe, then four bundles per minute would be even safer!

A Merlin powered Handley Page Halifax heavy bomber

Several minutes and a few dozen bundles of Window later the extension tube became hopelessly blocked and ceased to function.

Tapping the radio operator on the head to sort out the blockage, the engineer was alarmed to see him disconnect the blocked tube from the fuselage floor, resulting in thousands of reflective strips of Window filling their respective crew positions.

The resultant chaos failed to impress the skipper as the radio became inoperative and sparked merrily, and his engineers log, packed lunch and several other vital bits of equipment vanished down the gaping hole in the floor as they groped around blindly in the pitch darkness. This was compounded by the wireless operator accidentally inflating his 'Mae West' life jacket which rendered him immobile!

On their safe return the hasty excuse was that they flew through a cloud of another aircraft's Window which explained their Christmas tree appearance with shiny metallic ribbons adorning all projections on the outside of the aircraft. What was a great deal harder to explain to the Flight Commander was how there was even more on the *inside!*

~ ~ ~ ~

Commandments of Helicopter Flight.

He who inspecteth not his aircraft, giveth his angels cause to concern him.

Hallowed is thy airflow across thy disc, restoring thine Transational Lift.

Let infinite discretion govern thy movement near the ground, for vast is the area of destruction.

Blessed is he who strives to retain his standards, for without them he shall surely perish.

Thou shall maintain thy speed whilst between ten and four hundred feet, lest the earth rise and smite thee.

Thou shalt not let thy confidence exceed thine ability, for broad is the road to destruction.

He that doeth his approach, and alloweth the wind to turneth behind him shall surely make restitution.

He who alloweth his tail rotor to catch in the thorns curseth his children's children.

Observe thou this parable, lest on the morrow thy friends mourn thee!

~ ~ ~ ~

During the war Catalinas were often operated from the strip at Broome as air sea rescue cover for other aircraft. Near the airport building at one end of the strip were the local toilets with three entrances with large signs over the doors, OFFICERS, NCOS, GENTLEMEN. They're still trying to work out who was which.

~ ~ ~ ~

A great yarn doing the rounds at the 1997 Melbourne GP, courtesy of *Motorsport News*, was one of the visiting media contingent describing an 'experience' aboard his 747 enroute Singapore to Melbourne:

I was just beginning to enjoy a Scotch when suddenly the cabin address system crackled and the pilot began to talk, sounding as smooth and cool as Chuck Yeager in a flat spin – a voice which all airline pilots seem to have perfected.

"Good evening ladies and gentlemen," he said. "Sorry to disturb your sleep, but I thought you might like to know that we have a little problem with the electrical system of the plane. We've lost all power to the controls. The engines are working fine but if we hit turbulence we will probably go completely out of control and plummet to the ground or into the water – it doesn't matter much. You're going to be dead. So if you would like to use your mobile phones to call your nearest and dearest please feel free to do so. It is not going to make any difference with the navigation system.

Normally we would suggest that you adopt the crash position but really that is totally pointless, so sit back, relax and enjoy the inflight programming. I'll see if we can get the cabin staff to lay on another drinks service."

There was a stunned silence in the cabin.

And then, at the back of the plane there was a kerfuffle. A gorgeous woman rose from her seat and began tearing off her clothes.

"Before I die I want to be loved," she said. "I need a *real* man to make me feel like a *real* woman!"

A couple of seats in front of her, a man got up, tore off his shirt and threw it at her.

"Hear love," he said in a broad Australian accent. "You can iron this ..."

~ ~ ~ ~

This story involves an Australian Airlines Boeing 727 on approach to ASCB, in touch with Canberra Approach, and also on downwind leg. The weather is fine, very hot, with a little cloud. The Canberra district is in drought, with lawns and gardens in desperate need of rain. The 727 is advised of crossing traffic, with the conversation going something like this:

APPROACH: "Tango Bravo Golf, traffic is an Air Force Mystere, your 11 o'clock position, report sighting!"

VH-TBG: "Ahh ... negative sighting, confirm 11 o'clock!"

APPROACH: "Affirmative, your 11 o'clock now moving your 12 o'clock, approximately four miles"

VH-TBG: "Ahh ... negative sighting, all we have is a large cloud in our 12 o'clock, with a little rain."

APPROACH: "Tango Bravo Golf, forget the Mystere, grab the cloud, bring it over to my place, my front lawn is nearly history!"

RAAF Mystere

Bill Lines

~ ~ ~ ~

The incident below involves an IFR Beech Queen Air, VH-AEQ, Melbourne Approach, and Melbourne Tower.

MELBOURNE APPROACH: "Alpha Echo Quebec, clear for final, contact Melbourne Tower on 120 decimal 5 at Epping."

VH-AEQ: "Alpha Echo Quebec."

Then ...

VH-AEQ: "Melbourne Tower, Alpha Echo Quebec, final, left three thousand."

(No reply ... for thirty seconds or so ... then ...)

VH-AEQ: "Melbourne Tower, Alpha Echo Quebec, we're on final, left three thousand" followed by ...

MELBOURNE TWR: "Err ... Alpha Echo Quebec, this is the Tower Security Guard, the man you're looking for, the Tower Controller, is downstairs answering a call of nature!"

VH-AEQ: " Well tell him I'm landing anyway!"

VH-AEQ was in fact shooting an ILS approach for Melbourne Airport after a flight from Sydney.

~ ~ ~ ~

During the '60s a flightcrew licensing clerk with the then Department of Civil Aviation in Australia issued Certificates of Validation to visiting overseas pilots wishing to fly VH registered aircraft in Australia. The certificates were endorsed with those aircraft on which the pilot was proficient and were also on the Australian civil register.

One day an American private pilot and his wife came into the Sydney office after arriving from the States that morning, eager to pick up a Certificate of Validation. After studying and passing the Air Legislation examination, he proceeded to check his logbook for compatible aircraft to be endorsed on his certificate. The usual ones were present – Super Cub, Tri-Pacer, Cherokee, Cessna 150, 172, 180, and 182. He insisted that the floatplane endorsement also be included on the certificate. He then asked for names and telephone numbers of operators so that he could hire a floatplane and start planning their flying holiday 'downunder'.

The issuer explained that there were only a small number of floatplanes in Australia and doubted if any were available for hire in the Sydney area. A puzzled look appeared on both their faces, surely a country as big as Australia with all those lakes would have numerous floatplane operators willing to hire out amphibian aircraft.

What lakes were they talking about? Then the penny dropped, they were talking about the dry salt lakes in Central Australia. He then had the job of explaining as gently as possible, that the lakes they were referring to were salt lakes and, worse still they were usually very, very dry.

It appears that he was a keen floatplane pilot and had spent many hours touring the lake districts of the USA and Canada flying all sorts of amphibian aircraft. Looking for new areas to explore they came across a map of Australia showing, you guessed it, all the lakes of inland Australia. Assuming that all lakes were filled with water they saved up their spare cents and headed 'downunder' for a new adventure flying from east to west via our great lakes.

~ ~ ~ ~

A reader of *Australian Aviation* magazine from Timaru on the spectacular South Island of New Zealand recently heard Qantas 65 talking with Christchurch Control.

QANTAS 65: "Ahh ... Control, Qantas 65, any speed restrictions for us?"

CONTROL: "Qantas 65 ... that's no. Use warp factor five, or just go into hyperdrive!"

~ ~ ~ ~

A young controller was apparently handling the Essendon (near Melbourne) ATIS recording for the first time. This is how it went.

ATIS: "Essendon Terminal Information Charlie" ... (with the female controller providing all the details of runways in use, wind, QNH, temperature, cloud cover, reminding all aircraft that on first contact with Essendon to notify receipt of Information Charlie. She then screamed:

"I did it!"

~ ~ ~ ~

It's perhaps difficult to get shot down when you are already on the ground though this did happen to one unlucky lad in Britain during WW2.

an RAF base was busy night and day with instrument flight training using a large Nissen hut crammed with gyratint Link trainers. With strict blackout orders enforced, the entire base is nicely blacked out except for one little window where the curtain has come temporarily adrift, just letting a slither of light into the darkness beyond.

As was wont to happen, a lone Ju 88 night intruder was ambling about hoping to get lucky and find a target of opportunity. In those days, RAF night fighters were visual only and about as useful as ice in an igloo. Anyway, the Luftwaffe pilot sees this one pin prick of light and knowing that he is in the vicinity of a large RAF base begins to execute a strafing run.

By a thousand to one chance a single 20mm shell hits one of the little Links killing its pilot in the process. To our knowledge this is the only record of anybody being killed in a simulator and highlights the fickle luck that goes with war. Wrong place, wrong time ...

Link Trainer

~ ~ ~ ~

A story involving the Australian Airlines 727 VH-TBG. It was approaching Adelaide and changing from Adelaide Control to Approach.

VH-TBG: "Approach, Tango Bravo Golf, 80 DME Adelaide on the 088 omni and I'm on top with Juliet."

APPROACH: "You lucky bugger!"

~ ~ ~ ~

How many of you know the origin of the term "Plastic Parrot" for the CT-4 aircraft? It's high time a few stories surrounding the origin of the name surfaced.

One classic "parrot" story originates from an early CT-4A stream navigation exercise to RAAF Wagga, in south west NSW, that the staff and students at 1 Flying Training School (1FTS) conducted as part of their Dual Navigation Syllabus.

Melbourne FTS (Flight Service) was quite busy passing traffic (ah ... those were the days!) relating to the lengthy and indeed increasing stream of CT-4s to the other aircraft on frequency. Somewhere west of Wangaratta, a civil aircraft, flown by what sounded like a rather elderly gentleman was going to great lengths to ensure he had copied all the traffic. When informed by Melbourne FTS that there was a stream of twenty of so CT-4s tracking northbound to The Rock and Wagga, the old guy's reply was:

VH-NIT: "Ah Melbourne ... November India Tango, if I see that many CT-4s at once, they'll just look like a big flock of parrots!"

Before Melbourne FTS could get a word in, a member of the 1FTS staff renowned for his quick wit on the radio came back with this great reply:

UNIDENT: "I'd rather have a big flock of parrots than one big galah!"

RIP the Plastic Parrot!

CT-4A 'Plastic Parrot'

~ ~ ~ ~

Mike button accidentally depressed, aerobatic single engined aircraft involved.

PILOT: "I fell outa another damned roll, am I ever screwed up?"

TOWER: "Aircraft that just transmitted the obscene radio call, this is Austin Tower, say to me your callsign!"

PILOT: "Listen Tower, I'm not that screwed up!"

~ ~ ~ ~

An Ansett 767 flying the Melbourne/Perth route as AN1169, callsign "Romeo Mike Delta", rotates from RW16 at Melbourne. In doing so, it collides with a fox on the runway, who is moving too slow to get out of the heavy jet's way. The Captain, a Canadian with a big sense of humour, politely informs the Melbourne Departures they may have a carcass on their runway.

Later on, on frequency 125.7, Melbourne Control confirms with Romeo Mike Delta that he did in fact collide with and kill a fox during the strike. It brought this response from the fun loving Canadian skipper:

"Ok ... save the tail will ya, I'll pick it up on the way back for my car aerial!"

~ ~ ~ ~

A conversation heard between Speedbird 11, a British Airways 747, and Auckland (Oceanic) Radio.

SPEEDBIRD 11: "Auckland, Speedbird 11, we're at position RELIK at 1434, and request to be able to operate in the block, non standard Flight Level 295 through to Flight Level four two zero".

AUCKLAND: "Speedbird one one, this is Auckland, can I have your selcal code and your groundspeed?"

SPEEDBIRD 11: "Selcal is BRAVO PAPA CHARLIE DELTA, and our groundspeed is 573kt".

AUCKLAND: "Ahh ... Speedbird one one, for your information, and your passengers, the FA Cup final is still going. It's One All, and is now going to extra time!"

SPEEDBIRD 11: "Ohh ... there'll be penalties then, ohh ... that's marvellous!"

~ ~ ~ ~

Honolulu was heard selcaling Qantas 11. The Qantas aircraft was a Boeing 747-400, nonstop from Sydney to Honolulu.

HONOLULU: "Qantas one-one, Honolulu, call your company please, call Qantas Operations now."

QANTAS 11: "Roger that Honolulu, we'll call now."

QANTAS 11: "Qantas Sydney, this is Qantas 11."

QANTAS OPS: "Yes, your satellite message stated you have a bad odour in the aircraft. Please advise status."

QANTAS 11: "Qantas Ops, Qantas 11 ... it's at its worst around the galley and cockpit area."

QANTAS OPS: "Is it affecting First Class and Business?"

QANTAS 11: "No, but it's real bad in the cockpit!"

QANTAS OPS: "Regardless of that Qantas 11, have the Flight Services Director hand out Pax (passenger) Comment Cards to those areas, and return them as soon as possible to our Qantas Public Relations people. Our first priority is making sure no passenger was compromised regarding comfort!"

Several animals were placed in the forward hold for quarantine reasons. The odour resulted from their location.

Normally, the animals would have been carried in an aft hold, which would not affect comfort of crew or passengers. To their credit, Qantas apparently did everything in their power to make up for any unpleasantness caused by the odour, and all passengers went away satisfied. Nice to see they care so much about their customers.

~ ~ ~ ~

Speedbird one zero (British Airways flight 10) was departing just ahead of Qantas 5 out of Sydney when the following conversation took place.

DEPARTURES: "Speedbird 10, have you got thirty seconds?"

BA 10: "Speedbird ten, yes ... go ahead."

DEPARTURES: "My daughter is on board your aircraft today and I was wondering if you could look after her for me please?"

BA 10: "No problem ... her name please?"

DEPARTURES: (Passed the name of his daughter.)

BA 10: "We'll look after her."

DEPARTURES: "Thanks very much for that!" After a short time, the Controller vectored Qantas 5 into a right turn, which was acknowledged. Then ...

QANTAS 5: "What's the young lady doing flying with the opposition anyway?"

DEPARTURES: "I thought both your airlines are one and the same these days?"

QANTAS 5: "Not quite, they only own 25% of us."

BA 10: "But we're much better value aren't we?"

DEPARTURES: "Yeah, it was marginally cheaper with British!"

QANTAS 5: "Okay, I can relate to that."

~ ~ ~ ~

The following concerns an Ansett A320 being captained by an American pilot.

ANSETT A320: "Hotel Yankee Golf ready."

TOWER: "Hotel Yankee Golf, there's a little tacker on short final, line up behind that aircraft."

ANSETT A320: "A little tacker? What the hell's a little tacker, Tower?"

On another occasion, another A320 was again taxying for a Runway 23 departure, with another pilot on board also from the USA. The scenario was almost identical to our last story, but with a different third player ... Qantas!

ANSETT A320: "Hotel Yankee India ready."

TOWER: "Hotel Yankee India, a Qantas Joey on final, line up behind that aircraft."

ANSETT A320: "What's a Joey?"

VH-TAK: "Some more training's definitely required before you can be a real Aussie Ansett!"

The "Joey", of course, was the newly painted Qantas 737 VH-TAK, which was the first of the former Australian 737s to be painted in the Qantas livery.

~ ~ ~ ~

This story involves a well known Australian airline, inaugurating a new jet service on one of its regularly flown routes. To celebrate, a number of VIPs and media were invited along to take the first flight. Along the way ... yes ... the old chestnut ... the cabin address microphone accidentally keyed, when it was only intended to call the flight attendant servicing the flight deck.

"Okay honey ... how about two cups of coffee and lots of love?"

The female flight attendant heard the message, but so did everyone else in the main passenger cabin! Then the lady so "eloquently called" raced forward to the flight deck, obviously to stun the Captain with news of the gigantic error. Just as she opened the flight deck door, a VIP yelled loudly ... "Hey honey, you forgot the coffee!"

~ ~ ~ ~

Jack is from Bellerive in Tasmania.

Some years ago, he was approaching Tasmania's Cambridge Airport flying a Lake Buccaneer amphibian aircraft, a beautiful aeroplane.

Now Jack loves a joke as much as the next guy, especially when he can put one of them over on his friendly local neighbourhood air traffic controller.

He decided to go nautical. Something like this:

A/C: "Hobart, this is Lake Buccaneer Echo Tango X-ray, making landfall at Colebrook, at two hundred fathoms. Request clearance to enter the estuary for berthing at Cambridge."

There was silence. He had them! But then it happened, out of the ether ...

HOBART TWR: "Vessel Echo Tango X-ray, proceed to Richmond at present depth and await pilot!"

Jack reckons he's given up ... you just can't win against those quick thinking ATCs.

Lake Buccaneer

~ ~ ~ ~

Scene: Transit flight of four UH-1H helicopters, Mildura to Griffith. They are in loose formation, with all crews very noticeably bored.

UNKNOWN A/C: "Fives in."

Minutes pass in boredom.

UNKNOWN A/C: "Fives in."

Now, a cheesed off flight leader: "Will you guys down the back cut out the smart talk!"

A CHEEKY UNKNOWN A/C: "Fives in!"

LEAD AIRCRAFT: "Look, I'm sick of you guys with all the idle chatter, so shut up!"

A plea from the rear aircraft: "One from four, it's honestly not us. If you look out your right side, you'll see that FIVE IS IN!"

To which a now very inquisitive crew immediately glared, only to spy one very innocent looking Caribou, stealing into formation. Five was in!

~ ~ ~ ~

This story circa 1975, during flood relief operations in NSW:

FIRST A/C: "Do you know where you are Jack?"

SECOND A/C: "No, but I haven't crossed enough water to be outside Australia yet!"

~ ~ ~ ~

VH-WWH and VH-WWL are two floatplanes.

VH-WWL: "Whiskey Whiskey Hotel, this is Whiskey Whiskey Lima, OK, we have you visual."

VH-WWH: "We also have you visual, and you sure are ugly!"

~ ~ ~ ~

And again, this time circa 1978:

VH-MTR: "Melbourne, this is Mike Tango Romeo, departed Moorabbin 2028, six POB" (persons on board).

MELBOURNE: "Mike Tango Romeo, this is Melbourne. Moorabbin Briefing advises that you have left one of your passengers behind."

VH-MTR: "Melbourne, this is Mike Tango Romeo, roger, POB five!"

~ ~ ~ ~

On the flightdeck of an SAA Boeing 747SP flying Perth to Sydney on the final leg of its Indian Ocean service to Australia somewhere over the Great Australian Bight when the following words broke the stony silence:

UNIDENTIFIED AC: "I'm f.. ..g bored!"

ADELAIDE CTL: "That last aircraft station that just transmitted, identify callsign thanks!"

UNIDENTIFIED AC: "Sorry mate ... I'm not that f.. ..g bored!"

~ ~ ~ ~

This true story comes from Hong Kong, and involves a DC-10 on approach for Kai-Tak.

The Captain announced on the PA to his passengers that they were 170 miles from Hong Kong with a groundspeed of 540mph. They would soon be commencing descent and would be on the ground in 35 minutes. A quick bit of mental arithmetic by one passenger gave a different answer. He reasoned that 170 miles at 9 miles a minute was a bit under 20 minutes, not 35! Feeling 'rather superior', the plucky passenger sent a note written on a beer glass coaster, to this effect, drawing the Skipper's attention to the 'error'.

Back came the Aussie Captain's reply, conveyed by a respectful hostess: "Thank you for your note, but I have to stick by my arithmetic. How would you like to land this baby at 540 miles per hour?"

~ ~ ~ ~

A British Airways 747, callsign 'Speedbird one-one', was on final approach to Tullamarine's runway 27.

BA11: "Melbourne Approach, what altitude is that airship we can see just above the city skyline?" (just prior to changing to tower frequency)

APPROACH: "Speedbird-one-one, I have no knowledge of any airship in the area, so I can't help you there. I don't have him on radar either."

BA11: "I think it's one of Bondy's. Don't worry, there probably aren't enough tinnies on board to give you a radar reading!"

~ ~ ~ ~

An Australian Airlines 727 called Sydney Approach and the story goes like this:

VH-TBR: "Sydney Approach, Tango Bravo Romeo, on descent to seven thousand with 'tango'."

APPROACH: "Tango Bravo Romeo, five thousand."

VH-TBR: "Tango Bravo Romeo, five thousand."

APPROACH: "Tango Bravo Romeo, will advise runway shortly."

About 30 seconds later ...

APPROACH: "Tango Bravo Romeo, runway two five."

VH-TBR (in a very happy voice): "Oh goodie!"

~ ~ ~ ~

Overheard were Hornets being tested callsign 'Peregrine 632' on Melbourne Control frequency 123, when the jets were in the Mansfield area, near Eilden Weir, tracking for Avalon, which is to the southwest of Melbourne.

PEREGRINE 632: "Ah Control, Peregrine 632, my TACAN is unserviceable, could you give us a position fix please."

MELBOURNE CTL: "Roger Peregrine 632, you're five miles south of Mansfield, with seven-zero miles to run to Laverton."

Brief pause, then ...

MELBOURNE CTL: "Peregrine 632, what other Navaids are you equipped with?"

PEREGRINE 632: "None others except the Mark 1 EYEBALL."

(Perhaps this is a new, possibly super secret auxiliary Navaid, which could be said to be 'pilot integrated'.)

~ ~ ~ ~

The 1989 Australian pilots' strike provided excellent airband listening, what with the occasional snide remark made to our band of visiting replacement domestic airmen by air traffic controllers and other pilots! But all the interesting stuff isn't limited to aircrew and ATC staff. Some good one-liners have come from the mouths of passengers, such as overseas visitor Professor Bernd Nothofer.

The Professor had the privilege of riding as a passenger on board a stand-in RAAF C-130 Hercules aircraft from Sydney to Canberra. The learned gentleman had just arrived in Australia ex Frankfurt and was bound for a language conference in the national capital.

On board the cruising Hercules, a fellow passenger, an Australian, waited until the Professor had consumed the contents of his RAAF handout luncheon box and then said with a sneer ... "Well, what did you think of that?"

Not quite what the enquirer was expecting, the Professor replied earnestly. "Actually, it's much better than the breakfast we had on the way out here from Singapore!"

~ ~ ~ ~

A glider had been operating locally at Broken Hill and it reported inbound and was passed the surface conditions including wind, which was southerly at 5 knots.

GLIDER: "Are you sure about that wind? At the rubbish dump, it's westerly at 10 knots judging by the smoke."

BROKEN HILL: "All... right, are you going to land at the airport, or the rubbish dump?"

~ ~ ~ ~

This conversation was heard between a PanAm 737 and the Berlin Centre a few years ago.

The PanAm clipper 737 was inbound to Berlin Tegal down the central corridor, where the Americans generally used to fly as fast as possible, around 300 knots. More conservative pilots, particularly from the UK, used to prefer 250 knots. A Dan Air aircraft had the PanAm 737 behind it and the clipper was quickly catching them up.

BERLIN CTR (agitated): "Clipper 258, reduce speed now please!"

PANAM 258: "Ahh, clipper two-farve-eight, roger that."

BERLIN CTR: "Ahh ... clipper 258, what speed do you reduce to?"

PANAM 258: "Eventually zero ... I think!"

Typical quick-witted, laid-back American.

~ ~ ~ ~

It was an extra cold winter at Uranquinty in 1953 and a new rigger was doing a 400 hour inspection on a Wirraway.

Part of that job was to check the brake linings. They were almost down to the rivets and required replacement so he checked the maintenance handbook – parts to be returned to the manufacturer for relining! As the last Wirraway was delivered to the RAAF around 1946 he asked the Flt Sgt what I should do. "Why! It's a do it yourself job now". He replaced the linings and fitted them to the aircraft, and set the clearances to requirement.

Anyone who has worked on the Wirraway would know how difficult and time consuming this particular job can be. It was a regular thing to see a Wirraway being towed on its A frame to burn in new shoes. This aircraft was no exception.

A taxi test was then called for. A flying instructor who will remain nameless was organised. He climbed into the drivers seat with the Flt Sgt going along for the ride. Off they went out into the paddock. Put on the power to almost takeoff speed and applied the brakes to bring the aircraft to a standstill. Turn around and off again. This went on for some time with the pilot finally returning to the flightline outside the maintenance hangar.

The rigger climbed onto the port wing to ask the Flt Sgt how things were going. Before he had time to get an answer to the question, off went the aircraft once again, this time with three people on board, the pilot, the Flt Sgt and the rigger on the port side walkway.

Holding on for grim death, knuckles white holding onto the rear cockpit rails, assisted by the Flt Sgt holding him by the collar of his overalls. Then came the exciting part, the pilot continued taxi testing as before, throttle wide open almost to flying speed and braking to a halt, then around once more to repeat the procedure. The rigger had no idea how many near flights he had but was very glad when the pilot decided that he should return to the hangar line, switched off and turned round to see the rigger beside the rear seat passenger.

The pilot said to him "You got up there quick" all he could reply was that "he had not left yet". The look on the pilot's face was priceless.

All thought it was a great joke, with two exceptions, the 19 year old rigger and the rear seat passenger.

~ ~ ~ ~

VH-WDF: "Good morning to everyone travelling with Eastern Australia to Williamtown and further afield, this is your Captain. Our flight time to Williamtown from our present position 30 nautical miles north of Sydney will be approximately 35 minutes, and there is a little storm activity between here and Newcastle which we'll do our best to avoid and make your ride as comfortable as we possibly can. If you look in the pouch mounted behind the seat in front of you, you'll see a yellow card listing all the safety features of the Bandeirante aircraft, and the location of all of the emergency exists. We trust your flight is pleasant, and you will come fly with Eastern Australia again soon."

ATC: "Ahh ... Whiskey Delta Foxtrot, this is Sydney Departures, copied all thanks very much, but what colour is the little card I'm supposed to look at in front of me? I've looked everywhere and I just can't seem to see it! Oh ... and do we get any light meal?"

This is another situation where words totally escaped the aircrew! I think they were sliding well down in their padded seats trying to find somewhere decent to hide!

~ ~ ~ ~

The following story involves an Ansett 727 chartered by the Variety Club for the purpose of taking disabled children for a joyride. The aircraft was painted by Rolf Harris.

FLIGHT SERVICE: "Ahhh, Romeo Mike Lima, Sydney, what is the purpose of this trip?"

VH-RML: "Sydney, Romeo Mike Lima ... Variety Club chartered the aeroplane, Mobil paid for all the fuel, and the exterior of the aeroplane is painted in all jazzy colours by Rolf Harris and others."

FLIGHT SERVICE: "Oh excellent!"

Thirty seconds later, on this rather dreary rainy day in Sydney, the FIS operator came back with this schmuck comment, followed by a great reply:

FLIGHT SERVICE: "Romeo Mike Lima, I hope that paint's waterproof!"

VH-RML: "Course it is ... they painted it all over the Dulux Hi Gloss!"

The Ansett Variety Club 727.

Keith Myers

~ ~ ~ ~

Scenario: A training flight in a Cessna 152 south of Jandakot (Perth), with the student a bubbly young female with all of five hours in her log book and an instructor named Mike.

ATC: The Jandakot Tower Controller knows Mike well and is very aware he is the instructor on board VH-RAC.

The training session concludes and the aircraft turns inbound to Jandakot.

VH-RAC: "Jandakot Tower, Romeo Alpha Charlie, Forrestdale Lake, one-thousand-five-hundred and I've just received Mike!"

TOWER (unable to control his laughter ...) "Romeo Alpha Charlie, overfly at One-thousand-five-hundred and by the way ... we do sympathise!"

The young lady did realise what she had just said and during her next radio transmission was also unable to keep her laughter at bay, along with Mike chuckling in the background. Who said flying was always a serious business?

~ ~ ~ ~

This was heard from Coolangatta.

VH-ANE: "Coolangatta Tower, Alpha November Echo, flight 99 for Sydney, received Hotel, request taxi clearance."

TOWER: "Alpha November Echo, clear to taxi, time 56, when ready, clear for take-off! Make a right turn."

VH-ANE: "Wow! That's service for you!"

TOWER: "If I were you, I'd demand the same thing in Sydney."

VH-ANE: "That's a joke!"

~ ~ ~ ~

A friendly rivalry exists between pilots flying different types of aircraft in the RAAF.

For example, transport pilots are often known as Trash Haulers or Trashies, while those who fly fighters readily answer to the name Knuckle Head or Knuck. Also, for some obscure reason, fighter pilots have a reputation for liking jelly beans. This probably had its origins in WW2 where the convenient size and high glucose content of jelly beans made them ideal for giving pilots an energy boost during long escort missions.

Most people can take a joke, even when it is directed at them, but there are always those who never see the funny side of anything. Some years ago, a Trashie crew was supporting a flight of fighters on their way through Townsville to Darwin for exercises. They were in Base Operations studying the flightplan when a Wing Commander and Commanding Officer of the squadron concerned came in to submit their flight plans.

The big bowl of jelly beans which someone had placed on the flight planning table quickly caught his eye. But the man had no sense of humour and the Operations Officer, a Flight Lieutenant of many years service, duly suffered his indignation. The Wing Commander saw no fun in the gesture and went on at length about how professional fighter pilots really were and just how sick and tired he was of Knuckle Head and jelly bean jokes. He then took back his copies of the flight plans, turned and stormed out.

As the door slammed shut, the Flight Lieutenant, who until now had remained silent, looked at us and casually stated: "Wait till he opens his lunch box at thirty thousand feet!"

~ ~ ~ ~

In 1977, an Australian was in command of a Cessna 182RG, registered N736AC, 18 miles north of San Francisco, bound for Lake Tahoe, and unlike what happens in Australia, he was allowed to climb to 13,000 feet (no flight levels in the USA below FL 180) without oxygen to clear the 10,000 feet Sierra Nevada mountain range. Radio traffic was as depicted here:

OAKLAND: "Six Alpha Charlie, report 8000."

N736AC (Geoff): "6 Alpha Charlie."

OAKLAND: "6 Alpha Charlie, Oakland Radio, you a limey?"

N736AC: "Negative sir, Australian."

OAKLAND: "An Aussie! Well now! My daddy was in Brisbayne in the war!"

N736AC: "Oakland Radio, N736AC, great to hear, come and visit us sometime!"

OAKLAND: "I will. Have a nice day."

(Then 30 minutes later)

OAKLAND: "6 Alpha Charlie, Oakland, you still at 8000 on 047 degrees squawking 4368?"

N736AC: "Affirmative, SSR activated." (No SSRs in Australian light aircraft before 1981!)

OAKLAND: "Now boy, turn a mite quickly to 270 degrees, as this Bonanza is coming up behind you. If you don't he's going to wax your backside and spit you out his ass!"

(Can you imagine an Australian controller saying that?)

~ ~ ~ ~

An old woman enquired of a female flight attendant ... "How can the pilot see where he's going ... it's so dark outside!"

With that, the sharp witted young flight attendant pointed out the elderly passenger's window, indicating the navigation light on the starboard wingtip.

"Do you see that green light out there madam?"

"Yes", said the old lady.

Then the FA pointed out through the port window on the other side of the aeroplane. "And do you see that red light over there?"

"Yes", said the old lady.

"Well ... " said the flight attendant ... "as long as the pilot keeps the aeroplane between those two lights, we're okay!"

~ ~ ~ ~

Tango Bravo November (Boeing 727) was landing on RW07 at Sydney, having to cope with a 25 knot (45km/h) tailwind at 500 feet, and then ... they hit a plague of moths at 300 feet above ground level.

The following conversation took place after the aircraft landed, on the Sydney Ground frequency 121.7 MHz.

VH-TBN: "Err ... Ground ... we hit a moth plague at 300 feet or so on final for runway zero-seven. It was so thick that the windscreen is now opaque. Do you know why they're here?" (the crewmember on the radio very obviously shocked).

GROUND: "No, not really ... we've just been discussing it."

VH-TBN: "They're normally around Canberra!"

GROUND: "I won't tell anyone you just said that!"

~ ~ ~ ~

Geoff Westbury of Newcastle was kind enough to share his experience of April 18 1975 when he set an FAI recognised Australian light aircraft altitude record in a Piper PA-28-180, registered VH-RNP. The altitude attained was 18,300 feet and would have gone a little higher had not the oxygen supply run out ahead of the expected time frame.

Geoff departed Maitland for Scone and Quirindi on climb to FL 200, then returning to the Royal Newcastle Aero Club at Maitland. Approaching 10,000 feet, Sydney (121.6) asked him to call Control on 130.1 for onwards clearance. Between Scone and Murrurundi, climbing through FL 177, the following occurred:

EAST WEST F27: "Sydney, this is Echo Whiskey Delta, flight number 136, requesting clearance via flight planned route at FL 160."

SYDNEY FIS: "Echo Whiskey Delta, Sydney, you're cleared, climb to and maintain flight level 160, report at time 26."

(Nothing follows for one minute, then ...)

SYDNEY FIS: "Echo Whiskey Delta, Sydney, traffic. Romeo November Papa, aircraft type PA-28 oblique 180 at present on climb through flight level 177 for flight level 200, one o'clock to you, five miles."

EWD: "Sydney, Echo Whiskey Delta, say again aircraft type!" (unbelieving)

SYDNEY: "Echo Whiskey Delta, Sydney, it's a Piper PA-28-180."

EWD: "Sydney, confirm you did say a PA-28-180?"

SYDNEY: (slightly detectable mirth) "Affirmative!"

(Thirty seconds now pass, then ...)

EWD: "Sydney, Echo Whiskey Delta, are you saying ... er ... confirming that the traffic level 177 is a Piper Cherokee?"

SYDNEY (suppressed mirth): "Affirmative, separation more than adequate."

EWD: "Sydney, Jesus ... er ... well now I've heard everything!"

The altitude record was mentioned that afternoon in the Sydney press, and the next edition of ACPA. Our hero Geoff would have loved to have heard the replay of the CVR on the flightdeck of the F27!

Piper PA-28-180 Cherokee similar to the one in the story

~ ~ ~ ~

The piece is from the day the Gulf War officially ended and involves Brisbane Control and an Ansett BAe 146, VH-JJY.

VH-JJY: “Ahh, Brisbane Control, Juliet, Juliet Yankee, would you be able to give us co-ordinates for Kuwait, and heading, from our present position please?”

BRIS CTL: “Err, Juliet Juliet Yankee, standby.” (This was a lady controller, and a cool one too!)

A short time later ...

BRIS CTL: “Juliet Juliet Yankee, Brisbane Control, your Kuwait co-ordinates are ... (rattling off a set of numbers too fast for Kel to write down).

Then came stunned silence, followed by ...

VH-JJY: “Err, Control, Juliet Juliet Yankee, gee ... thanks, err, if we divert now from ENTEC, we could be there by err, ... tomorrow this time! It’s only 7000 nautical miles!”

Not to be outdone ...

BRIS CTL: “Juliet Juliet Yankee, call Control on 123 decimal zero, then send me a postcard from the Gulf. See ya! (giggling)

~ ~ ~ ~

An interesting experience was encountered while flying LanChile to Santiago from Tahiti.

On approach to Santiago, the pilot came over the intercom, and told the cabin crew in Spanish to “prepare for the paper landing”. No one seemed to know what he meant, that is ... until finals. It would appear that the pilot decided to land the aircraft in a “slip” condition. With the crosswinds, and the rain in the area at the time, the landing didn’t go exactly according to plan. In theory it worked, but in practice, things went off the edge of the plate. The aircraft “crabbed”, nearly running off the runway.

Following the landing, the pilot apologised to his passengers, and then blamed the landing gear for the problems!

~ ~ ~ ~

The following situation went as follows, Brisbane Ground frequency, a 747-400 (Qantas 28) was taxying to park at Bay 4, while a Qantas 747SP, known to other crews as ‘Classic’, ‘Dinosaur’, ‘Coal Burner’, ‘Grey Line’, or ‘Analogue Antique’ was pushing back from Bay 3. The archaic Brisbane system of single taxiways meant that the SP was requested by Ground Control to push back far enough to allow Qantas 28 to enter Bay 4. Dutifully, the SP crew pushed back to behind Bay 5 to clear the approach for the arriving -400 aircraft.

Now, at this moment in time, the skipper of Qantas 28 begins to turn into his allotted bay a little too early to ensure his longer wingtip clearance from the smaller SP wing that was facing him. All flightdeck heads were turned SP-ward to ensure nothing touched what it shouldn’t, when the following incredibly funny remark was heard over the radio, the “culprit” most likely being from the crew of the 747SP.

“I really wouldn’t worry ... your wing tip’s already bent!”

~ ~ ~ ~

An F-111 from RAAF Amberley was operating in the Coolangatta control zone, testing the new radar.

Part of the routine involved "maximum rate turns". Meanwhile, an Australian Airlines 727 (VH-TBO) made its initial Tower contact with Coolie, quite unaware of the F-111, or what it was doing.

The controller transmitted ... "Buckshot triple-one, request your angle of bank and maximum 'g' forces."

Buckshot 111 replied with ... "85 degrees AOB, and we are pulling a max of 4.5gs."

A split second later, a crew member of the 727 queried the Tower controller with ... "what airline's that?"

Quick Tower reply: "Bob Hawke's!" (Aussie Prime Minister at the time)

RAAF F-111

~ ~ ~ ~

From the early seventies, and the days when Changi in Singapore was a military airport only, and Paya Leba was the international gateway.

A lot of charter operators worked out of Singapore through to Europe, undercutting the airline's passenger fares. Some of these operations were probably a little bit questionable as the following conversation shows:

TOWER: "Fly-by-night 789, line up!"

A Boeing 707 dutifully rolled onto the active runway and reported ready.

TOWER: "Fly-by-night 789, clear for ... err, Fly-by-night 789, let me ask ... have you paid your landing fees yet?"

707: "Err ... well ... err."

TOWER: "Fly-by-night 789, you take the next @#%* taxiway left, return to dispersal, and pay your fees!"

~ ~ ~ ~

From Canberra comes the tale of an Eastern Airlines Dash 8 departing ASCB for ASSY in the early afternoon.

After receiving takeoff clearance from Canberra Tower, the cheeky pilot acknowledged, adding ... "by the way ... we don't wish you any luck for this afternoon!"

TOWER: "And we don't wish you any luck either!"

An unusual air/ground conversation you might think! Actually, there was good reason for these aggressive remarks ... the Rugby League Grand Final between Canberra and Sydney club Penrith.

Penrith emerged victorious, so if that particular Dash 8 crew flew back into Canberra after the game that afternoon I dare say the exchanges on the radio would have been even less friendly!

~ ~ ~ ~

A young Lieutenant aboard HMAS *Melbourne*, on a beautiful autumn day early in 1975. The carrier was off Jervis Bay, carrying out the very boring task of three dimensionally mapping the warship's radar coverage area, using a chartered Learjet, flying back and forth at high altitude. Our Lieutenant visited the ATC Officer, who like the pilot aloft, was exquisitely bored. At the end of the sortie, the ATCO suggested the pilot might like to fly one carrier controlled approach letdown, culminating in a low level overshoot.

The pilot agreed and when the Lear was half a mile out, our Lieutenant made a dash down to the flightdeck to watch the overshooting Lear. There were only two young sailors on deck, both without shirts, enjoying the sun while painting a new runway centreline on the deck in yellow paint. Looking for some fun, our young Officer yelled "lookout" and pointed towards the now rapidly approaching Lear, which had landing lights blazing and flaps and gear down, approaching at 130 knots (240km/h). The sailors froze, jaws drooping, eyes wide, then they executed their own spectacular takeoff, scattering brushes and cans of paint along with a ghetto blaster radio in their haste to save their own lives! They cleared the landing area just in time to watch the sleek jet fly by the port beam and gracefully depart for home.

The Lieutenant received some mighty aggressive looks and as such decided to execute his own graceful departure! The boring day drifted on, with only the fresh trail of yellow footprints on the flightdeck as sole evidence of the moment of alarm for the hapless young sailors!

~ ~ ~ ~

The following classic occurred at Archerfield in Brisbane.

Whiskey Juliet Whiskey, a PA-60 Cheyenne was in conversation with Archerfield Ground.

VH-WJW: "Archerfield Ground, Whiskey Juliet Whiskey, request an area clear of other aircraft to test our radar equipment."

GROUND: "Whiskey Juliet Whiskey, I suggest the threshold of 04 Left as it's not currently active."

VH-WJW: "Roger, thanks."

GROUND: "And Whiskey Juliet Whiskey, point that thing away from me, I don't want to be sterilised."

VH-WJW: "I thought you did."

GROUND: "Might as well ... for all the use it is to me!"

~ ~ ~ ~

One afternoon, a guy had just been for a spin (literally!) as a passenger in a Pitts Special. After his pilot had thrown the Pitts around the sky doing snap rolls, loops, spins and the like, the duo returned in their Pitts to Bankstown for a RW29R arrival. The passenger then thumbed his press-to-talk and jokingly suggested to the pilot that a "snap roll would be great on final". Yes ... the old chestnut. The press-to-talk was the "tx" and not to "intercom". The passenger was blissfully unaware that the whole world heard his suggestion. On vacating the runway, the Tower came back with ... "November 722, although we could never have approved that manoeuvre it would have been *great!*"

~ ~ ~ ~

A couple had flown into San Francisco and were taking the shuttle down to Los Angeles.

The early model 737 was very sparsely equipped, only having open overhead lockers, and no babies cribs as found on the Qantas 747 they were to connect with. The lack of a suitable place for their son consequently resulted in him on their laps for the flight.

When they landed at LAX the pilot applied full braking and reverse thrust. The effect was dramatic to say the least, whilst they tried to keep their nine month old from being thrown into the forward bulkhead, the drink trolley broke loose from somewhere up the back and literally flew to the front of the aircraft. It hit the middle of the cockpit door perfectly. Luckily for the pilots this was locked. Bottles, cans, cups and packets of nuts were distributed over the entire front end of the cabin.

They asked one of the cabin staff, who was tidying up the mess, if this was normal procedure on landing in LA. She replied, "If you enjoyed that sir, you should come with us to San Diego, we get a bonus if we can take the first taxiway off the runway."

~ ~ ~ ~

Boeing 747 Qantas 4 had just departed from RW16 at Sydney and was on climb, but still monitoring the Tower frequency. And then ...

UNITED 815: "Aah, Sydney Tower, United 815 is taxying up to one-six." (Imagine this fellow with a very deep, slow southern drawl.)

TOWER: "United 815, Sydney Tower, line up runway one-six, hold position."

UNITED 815: "United 815 is rolling for one-six and will hold."

TOWER: "United 815, Sydney Tower, cancel SID, maintain runway heading, maintain three thousand, clear for takeoff."

(At this point in the proceedings, there is a considerable pause, but United 815 finally acknowledges.)

UNITED 815: "United 815, ah Sir, could you please explain to me how I can maintain three thousand when I'm still sitting here on the runway?"

(Total silence from Sydney Tower) then ...

TOWER: "United 815, Sydney Tower, after taking off, maintain the runway heading, climb to three thousand feet, when you reach that altitude, maintain until further advised. Clear for takeoff runway one-six."

(A garbled acknowledgement from United 815, then a voice from out of the blue, in a very noticeable Aussie accent: remember Qantas 4 still on frequency.)

UNIDENTIFIED A/C: "Do you come here often?"

(A lot of silence from EVERYONE follows now.)

~ ~ ~ ~

An Ansett 737 enroute to the Gold Coast from Melbourne, climbing to flight level 330.
MELBOURNE CONTROL: "Charlie Zulu Foxtrot, call when passing flight level 210."
VH-CZF: "Charlie Zulu Foxtrot, no problems."
MELBOURNE CONTROL: "You're working late tonight."
VH-CZF: "Yeah, going up to the Gold Cost. We'll be touching down about 12 past 12."
CONTROL: "That's late?"
VH-CZF: "Yeah ... that's what you get when you haven't got a flight engineer!"
CONTROL: "It figures."
VH-CZF: "Not to worry, I'll get home and watch a bit of the tennis."
CONTROL: "Who's your tip this year?"
VH-CZF: "The Pat Brat!"
CONTROL: "Charlie Zulu Foxtrot, report flight level, then call Control on 118.7."

~ ~ ~ ~

This amusing offering is a tale of security precautions gone wrong, wires crossed and tempers flared, but with a happy ending! (Happy endings are unusual these days when security misses a beat!)

The scene was Frankfurt Airport, West Germany, February 1985. The boarding lounge. A call was made to board a flight to Innsbruck, Austria, on a regional carrier. It was announced that as it is often the case at Frankfurt, passengers would be bussed out to the aircraft, where they would identify their luggage before boarding. The passengers all trooped onto the bus and arrived at the aircraft, which was being watched over by a sprinkling of armed soldiers and security men, all looking very earnest and ferocious. But there was not a suitcase in sight.

While the situation produced some very worried looks and comments from the passengers, it did not phase the employee of the host airline in Frankfurt, in the slightest! She directed the passengers all towards the steps of the aircraft to board. Several people asked her where the luggage was and these questions were met with that practised look of arrogance that airline employees the world over have mastered!

The obvious answer to her was that the luggage was on board already ... where else could it be? Most of the passengers accepted this, but one fellow, of Germanic origin, insisted that the passengers had been told that the luggage was to be identified and why has it been loaded?

While the eyes of the furious girl glazed over, he demanded (this time of the soldiers) to identify the luggage. Thereafter ensued a terrific argument between the girl, the soldiers, the baggage handlers and finally the hapless crew of the aircraft as to why the luggage was loaded, why it wasn't to be searched, who said anyway that it had to be searched and did they all have to stand out there in the cold and fight? These last sentiments were echoed by a chorus of passengers!! Gilbert and Sullivan at its best!

Eventually, the guys with the guns won and the luggage was unloaded, while the passenger who started the whole thing beamed self-righteously. As each item of baggage was identified, it was dutifully reloaded and 50 rather frozen passengers eventually boarded the aeroplane, which without further ado and with the blessings of the security conscious passenger departed.

~ ~ ~ ~